Lecture Notes in Applied and Computational Mechanics

Volume 37

Series Editors

Prof. Dr.-Ing. Friedrich Pfeiffer
Prof. Dr.-Ing. Peter Wriggers

Lecture Notes in Applied and Computational Mechanics

Edited by F. Pfeiffer and P. Wriggers

Further volumes of this series found on our homepage: springer.com

Dynamics and Balancing of Multibody Systems

Himanshu Chaudhary · Subir Kumar Saha

Springer

Dr. Himanshu Chaudhary
Dept. of Mech. Eng.
Malaviya National Institute
of Technology Jaipur
JLN Marg, Jaipur-302017
Rajasthan
India
himanshubhl@rediffmail.com

Dr. Subir Kumar Saha
Dept. of Mech. Eng.
IIT Delhi
Hauz Khas, New Delhi 110 016
India
saha@mech.iitd.ac.in
sahaiitd@yahoo.co.in

ISBN: 978-3-540-78178-3   e-ISBN: 978-3-540-78179-0

Lecture Notes in Applied and Computational Mechanics   ISSN 1613-7736

Library of Congress Control Number: 2008934651

Cover design: WMXDesign GmbH, Heidelberg

Printed on acid-free paper

9 8 7 8 6 5 4 3 2 1 0

springer.com

Dedicated to the

Rural People of India and The World

Preface

This book has evolved from the passionate desire of the authors in using the modern concepts of multibody dynamics for the design improvement of the machineries used in the rural sectors of India and The World. In this connection, the first author took up his doctoral research in 2003 whose findings have resulted in this book. It is expected that such developments will lead to a new research direction MuDRA, an acronym given by the authors to "Multibody Dynamics for Rural Applications." The way MuDRA is pronounced it means 'money' in many Indian languages. It is hoped that practicing MuDRA will save or generate money for the rural people either by saving energy consumption of their machines or making their products cheaper to manufacture, hence, generating more money for their livelihood.

In this book, the initial focus was to improve the dynamic behavior of carpet scrapping machines used to wash newly woven hand-knotted carpets of India. However, the concepts and methodologies presented in the book are equally applicable to non-rural machineries, be they robots or automobiles or something else. The dynamic modeling used in this book to compute the inertia-induced and constraint forces for the carpet scrapping machine is based on the concept of the decoupled natural orthogonal complement (DeNOC) matrices. The concept is originally proposed by the second author for the dynamics modeling and simulation of serial and parallel-type multibody systems, e.g., industrial robots, Stewart platform used as flight simulators, and others. As reported earlier, the use of the DeNOC matrices for the above systems has resulted in efficient and numerically stable computer algorithms. Hence, the attempt is made here to write the DeNOC matrices for the closed-loop carpet scrapping machine which does not have any special feature like serial or parallel. The above attempt has resulted in two-level recursive dynamic algorithms for general-type planar and spatial closed-loop multibody systems. The methodology is later exploited for the dynamic balancing of complex mechanical systems. Hence, the book has two distinct contributions, namely, in the domain of dynamic formulation and in balancing.

This book is useful for teaching in the graduate-level (Master's and Ph. D) courses specializing in Dynamics. In fact, the contents of the book can be taught over one or two-semesters. If two semesters are chosen to cover the materials of this book sufficient class projects must be assigned. In that case, the materials on dynamic modeling provided in Chapters 1-3 are to be covered during the first semester, whereas the concept of equimomental system and the balancing explained in Chapters 4-6 are to be covered in the second semester. The content of the book can also be utilized by practicing engineers engaged in the design of new machines, or want to modify an existing design for improved performances.

Organization of the Book

The book is organized into six chapters that cover dynamics and balancing of Multibody systems. To illustrate the basic ideas and procedures examples are used in each chapter.

Chapter 1: Introduction

Chapter 1 introduces the motivation and objectives of dynamics of multibody systems and the problem of balancing of inertia-induced forces in mechanisms. Along with the survey of literature related to these problems, the basic DAE (Differential Algebraic Equations) and ODE (Ordinary Differential Equations) formulations for multibody system modeling and simulation are also described.

Chapter 2: Dynamics of Open-loop Systems

In this chapter, first the six-dimensional twist and wrench vectors for spatial velocity and force are introduced to write the dynamic equations of motion concisely. The mathematical tool used for the dynamic formulation is the concept of decoupled natural orthogonal complement (DeNOC) matrices. The formulation of DeNOC matrices is explained for serial- and tree-type systems, which is extended in the following chapter for general closed-loop systems. The concepts are explained with several examples.

Chapter 3: Dynamics of Closed-loop Systems

The dynamic formulation for open-loop system presented in Chapter 2 is extended in this chapter for closed-loop systems. The concept of joint cutting methodology is adopted to convert closed-loops into open-loops. The removed constraints due to the cut joints are retrieved by introducing the Lagrange multipliers. Then, a novel systematic formulation of constraint wrenches for closed-loop systems, called *two-level recursive approach*, is introduced to compute the constraint moments and forces at the joints. The methodology and computation schemes are both illustrated with the aid of planar and spatial examples. The examples are so chosen in such a way that they demonstrate the advantages of the proposed methodology with respect to modeling complexity and computational efficiency.

Chapter 4: Equimomental Systems

This chapter presents the concept of *equimomental systems* of moving rigid bodies. Two and three point-mass models for planar motion and seven point-mass models for spatial motion are illustrated.

Chapter 5: Balancing of Planar Mechanisms

The recursive methods for the dynamic analysis of open-loop and closed-loop mechanical systems using DeNOC methodology, presented in Chapters 2 and 3, respectively, are one of the important contributions of this book. The other essential aspect is the solution of dynamic balancing of mechanisms. The balancing problem is formulated here as a general optimization problem, in which the concept of equimomental system introduced in Chapter 4 is used. Though the concept of equimomental system of point-masses is not new, its effective applications in mechanism design/synthesis is rather rare, especially for spatial mechanisms. Bringing this concept into light and showing its useful implementations in mechanism synthesis is one the salient feature of this book.

Chapter 6: Balancing of Spatial Mechanisms

The methodology developed and verified for planar mechanisms in Chapter 5 is extended in this chapter to spatial mechanisms. Similar to the balancing methodology proposed for planar mechanism, each link of a spatial mechanism is also treated using equimomental system of seven point-mass model. A spatial RSSR mechanism is balanced to illustrate the concept of balancing proposed here.

Appendix A: Coordinate Frames

In order to specify the geometric configuration of spatial mechanisms, body-fixed coordinate frames are defined. The selection of reference frames, representation of various vectors, transformation of vectors and matrices from one frame to other, etc. are explained in this appendix.

Appendix B: Topology Representation

In this appendix, the methodology to cut open a closed-loop system is presented based on the concept of cumulative degree of freedom used in graph theory.

Salient Feature of the Book

The contents of this book are well grounded in the current research trends of multibody dynamics and mechanisms synthesis. A modern and advanced methodology for dynamic modeling of open-loop and closed-loop systems is developed systematically. The underlying problems associated with the minimization of inertia forces are important to reduce noise, vibration, and wear of machines, etc., and to improve their fatigue life, etc. The salient features of the book are outlined below.

1) A critical review of methods for modeling and simulation of multibody systems, as well as the balancing of mechanisms, is provided.

2) The decoupled natural orthogonal complement (DeNOC) matrices are introduced for serial and tree-type mechanical systems for the purpose of dynamic modeling.

3) Concise forms of the dynamic equations of motion are derived using DeNOC matrices.

4) Recursive dynamic formulations of constraint wrench for both open- and closed-loop systems are developed.

5) A number of nontrivial examples are solved to illustrate effectiveness of the dynamic analyses using the proposed formulations.

6) Using the equimomental system of point-masses, the dynamic equations are appropriately reformulated.

7) A general mathematical optimization problem is formulated for the balancing of inertia-induced forces, both for planar and spatial mechanisms.

8) Considerable reductions in shaking force, shaking moment, and driving torque is shown for practical mechanisms.

Units and Notation

The international System of Units (SI) is used in the examples and problems solved in the book. Also, the boldface Latin/Greek lower case and upper case letters are used to denote vectors and matrices, respectively, whereas italic Latin/Greek letters with lower case are referred to scalars. In some cases there may be deviations, e.g., for zero vectors and matrices, and cross-product matrices. In such cases, they are defined as soon as they appear in the text.

Acknowledgments

We would like to thank all those who have helped us with their support and contributions in the preparation of this book. Special thanks are due to Indian Institute of Technology (IIT) Delhi where the first author did his Ph.D. and M.L.V. Textile and Engineering College Bhilwara for granting the study leave. The information provided on carpet scrapping machine by the Micro-model Laboratory of IIT Delhi is highly acknowledged. We also thank the people of Mechatronics Laboratory at IIT Delhi, particularly, Mr. Dharmendra Jaitly, Mr. Naresh Kamble, Mr. Ashish Mohan, Mr. Ali Rahmani, and others with whom we had many discussions about life and education that may have influenced the presentation of this book indirectly.

Special thanks are also due to our respective family members, Abhimanyu, Deeksha, Anjali and Madhu (with Himanshu Chaudhary), and Bulu and Esha (with Subir Kumar Saha) for their patience and understanding while this book was under preparation. In addition, we express our sincere gratitude to Springer Verlag, Germany and the anonymous reviewers for readily accepting the book for publication. The guidance of Heather King from Springer during the manuscript preparation is also acknowledged.

MNIT Jaipur   Himanshu Chaudhary
IIT Delhi   Subir Kumar Saha

Contents

1 Introduction

The Computer-Aided analyses of Multibody Mechanical Systems emerged as an important scientific tool in the Applied Mechanics field. This is possible due to improvements of the computer technology at the level of both hardware and software. Before the advent of computer-aided tools, the design of machines and their components was based on trial and error and graphical methods. In the last four decades, a great number of the methods and techniques has been proposed to analyze the dynamic behavior of systems consisting of multiple bodies. There is a great need to utilize those methods to produce cost effective components and machines with high reliability and durability. To achieve this goal, the analysis of complex mechanical systems which includes kinematic and dynamic analyses, synthesis, and optimization of forces must be viewed in an integrated manner. On the other hand, balancing is one of the crucial steps in the design of high speed machineries. It is a difficult problem due to the interaction amongst various dynamic quantities, e.g., shaking force, shaking moment, bearing reactions, and driving torques/forces. These interactions can, however, be considered in an optimization problem where various dynamic quantities are computed repeatedly. In this respect, this book presents a unified methodology for dynamic analysis and minimization of the inertia-induced forces occurring in high speed multiloop planar as well as spatial mechanisms based on the multibody system modeling approach.

1.1 Dynamics

A multibody system can be defined as a number of bodies interconnected with constraint elements such as joints, bearings, springs, dampers, actuators, and acted upon by external and internal forces as depicted in Fig. 1.1. Many mechanical systems like robots, heavy machinery, automobile suspensions and steering systems, machine tools, animal bodies, satellites, among others, are multibody systems [1, 2]. Other multibody systems are mechanisms and machines. With advances in computing facilities, computer-aided analysis of multibody dynamics emerges as a major tool for the analysis and design of multibody systems. It has been the topic of many research activities in the last four decades [1-79]. The main effort is usually focused on automatic derivation of the equations of motion, and numerical solution suited to predict the dynamic behavior of the systems under study. In this regard, there are two basic problems. The first problem is to determine the motion of a system from a set of applied forces. This is referred to as forward or direct dynamic problem. The second one is the problem of determining the forces required to produce a prescribed motion, as well as the constraint moments and forces, i.e., the reactions at the joints. Such problem is referred to as inverse dynamic problem. When the driving forces and torques are of only concern, as in the control of a system, the constraint forces are not determined in the inverse dynamics analysis. Figure 1.2 demonstrates the above definitions of the dynamic problems.

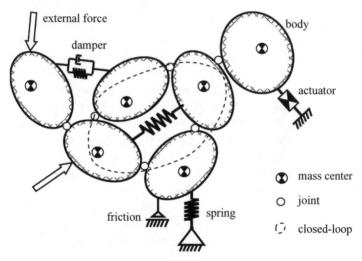

Fig. 1.1. A typical multibody system and its elements

Note that the forward dynamic analysis is essential to predict the system's behavior, whereas the inverse dynamics analysis is valuable for many reasons. The inverse dynamics has important applications in the control of a multibody system, where the driving forces to follow a specified trajectory are evaluated. Such a scheme is often referred to as the feed forward control law. Inverse dynamics also finds applications in designing the supports and joints, where the stresses induced in the machine components, fatigue characteristics, etc., are calculated using constraint forces.

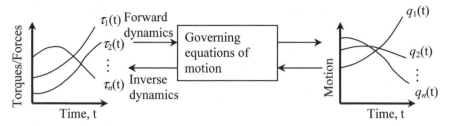

Fig. 1.2. Forward and inverse dynamic problems

1.2 Formulation of Dynamic Analysis

Dynamic analysis of multibody systems has been recognized as an important tool in design and simulation of complex mechanical systems, such as robots, vehicles, and mechanisms/machines. The study of rigid multibody systems is dated back to the period of Euler (1707-1783) and D'Alembert (1717-1783) in the eighteenth century [1, 2]. Their research gave birth to the celebrated Newton-Euler equations of motion for a rigid body, which are based on the Newton's laws of linear motion and the Euler's equations for rotational motion. A systematic analysis of constrained multibody systems was established by Lagrange (1736-1813), who analytically derived the generalized equations of motion by using energy concept. Lagrange's work was milestone in the development of classical mechanics, as it was derived independently of Newtonian mechanics. The Euler-Lagrange equations of motion for a constrained system have their roots in the classical work of D'Alembert and Lagrange on analytical mechanics, and in the work of Euler and Hamilton on variational calculus [3, 4]. A great number and variety of formulations on the governing equations of motion for constrained multibody systems have been developed since then [5-7]. The complexities in the formulation of kinematics and dynamics lie with the types of the coordinates selected, the basic principles of mechanics, i.e., Newton-Euler equations, D'Alembert principle, virtual work, Lagrange equations, the method selected for handling constraints, etc.[7-8]. With the advent of digital computing technology, computer-aided analysis of multibody dynamics has emerged as a major tool for the analysis of constrained mechanical systems. In the past four decades, researchers focused mainly on the automatic generation of the equations of motion and their solution strategies. Broadly, these formalisms can be classified into three categories, namely, the Newton-Euler (NE) approach, the Euler-Lagrange (EL) approach, and the Kane's method, as explained in Fig. 1.3.

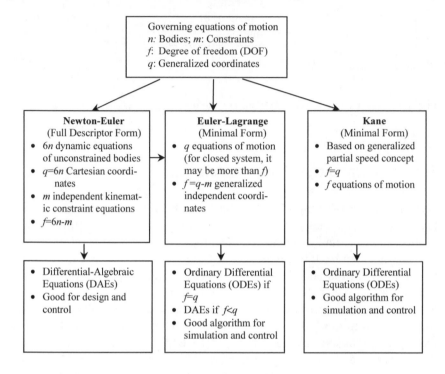

Fig. 1.3. Types of dynamic formulations

1.2.1 DAE vs. ODE

The NE approach [7-11] treats each rigid body separately, and physical vectorial quantities, such as velocity, acceleration, force, and torque are expressed in the absolute Cartesian coordinates. The EL [12-19] and Kane [20-25] approaches on the other hand look at the system as a whole, and use the concept of generalized coordinates and forces. Based on the NE or EL approach, the governing equations of motion of the mechanisms with closed-loops are of differential algebraic equations (DAE) type. Note that the EL equations can be derived in two forms: Lagrange's equations of the first kind or the equations in Cartesian coordinates with undetermined Lagrange multipliers, and of the second kind, i.e., the equations of motion in terms of the independent generalized coordinates. The EL equations of the first kind represent a set of DAEs while the second kind leads to a minimal set of ordinary differential equations (ODE) [2]. The Kane's equations are basically the "Lagrange's form of D'Alembert principle," as termed by Kane [21] himself.

Typically, index-3 DAEs, where the index of a DAE denotes the maximum number of times an equation in the DAE has to be differentiated to obtain the standard form of ODE [73], have sparse structure and can be solved efficiently using sparse matrix techniques and stiff time integration methods. Orlando et al. [13] first took the advantage of the index-3 DAEs and employed this technique in the commercial software ADAMS to simulate the constrained mechanical systems. Three different approaches have been proposed in the literature for integration of the DAEs [26-32]. They are the direct integration method, the constraint stabilization method [27-29], and the generalized coordinate partitioning method [31, 32]. In the direct integration method, numerical integration of the ordinary differential equations is used to integrate the index-1 DAE obtained by repeatedly differentiating the constraint equations of index-3 DAE. In this method, all the coordinates are integrated irrespective of their dependency. As the kinematic positions and velocity constraints are not taken into account, the integration of the index-1 DAE leads to solution drift and suffer from constraint violation. To overcome the drift and violation problem, Buamgartes's stabilization [27] widely used. Recent efforts have been reported that apply modern control theory to drive better stabilization forms, e.g., in [29]. Integrating all the generalized coordinates such as absolute coordinates is not useful as dependent coordinates can be computed from the independent ones which are equal to the degree of freedom (DOF) of the system. Hence, the methods of partitioning the independent and dependent coordinates are used in [31-32].

Irrespective of the substantial progress in numerical methods to solve the system of DAEs, they are considered difficult and less efficient than that of the ODEs. Therefore, a considerable amount of research effort has been spent to systematically reduce the dimension of the DAEs to a set of ODEs. This can be done by orthogonalization of the Jacobian matrix of the constraint equations. Huston and Passerello [33] introduced the first such method to reduce the dimension of the DAEs. They proposed vectors orthogonal to the direction of the constraint forces without giving explicit method for their determination. The inner product of these vectors with the equations of motion automatically eliminates the constraint forces. Later, Hemami and Weimer [34] coined the words 'orthogonal complement' in the modeling of nonholonomic systems. The proposed complement is nothing but the matrix form of the orthogonal vectors given by Huston and Passerello [33]. They indicated that the complement is not unique. In this regard, Kamman and Huston [22, 23] have illustrated the use of the zero eigenvalue theorem in obtaining the reduced set of equations when the governing equations of motion are derived using Kane's equations. Coordinate partitioning using LU decomposition [32], singular value decomposition [35, 36], methods based on Householder transformations [37, 38], and the Gram-Schmidt orthogonalization [39] are the representative examples of the methods of reducing the dimension of the dynamic equations of motion to a minimal set. Wampler et al. [40] proposed a procedure to reduce the dimension of the equations of motion based on finding the inverse of a matrix whose first few rows constitute the constraint Jacobian matrix and the remaining rows correspond to the augmented matrix that relates the generalized speeds. The elimination of the Lagrange multipliers, partitioning of the entire set of generalized coordinates into independent and dependent ones, better numerical integrator for the ODEs, and handling of system with redundant constraints or with varying DOF are some of advantages offered by these methods. Basically, independent base vectors of the tangent space of the constraint surface are calculated in these methods, and the projections are used to reduce the dimension of the equations of motion [41-44]. The tangential projection gives the constraint-free equations of motion, and the orthogonal projection allows one to determine of the constraint reactions. Based on the properties of the constraints, Hui et al. [45] proposed a method to automatically formulate the equations of motion of multibody systems with linearly independent constraints without finding the orthogonal complement matrix of the constraint Jacobian matrix.

1.2.2 Recursive formulations

Most of the recent improvements in multibody dynamic formulations have come from robotics and aerospace fields [46-63]. Stepanenko et al. [46] formulated the first recursive Newton-Euler equations for spatial open chains. In their formulation, the kinematics of the links is represented in fixed coordinate frame. The recursive Newton-Euler formulation proposed by Luh et al. [47] using moving coordinate frames seems to be the most efficient solution to inverse dynamic problem, at least for the general geometries of the serial manipulators with six degree of freedom. The computational requirement of this method varies linearly with the number of bodies, i.e., its computational complexity is $O(n)$, where n is the number of bodies. Hollerbach [48] provided a recursive form of the Lagrangain approach, and produced another efficient $O(n)$ algorithm, which was shown to be equivalent to the NE method [49]. The development of efficient inverse dynamic algorithms prompted Walker and Orin [50] to use it as the basis for their procedures for forward dynamic problem. It is an $O(n^3)$ method and they called it composite inertia method. Jain [51] thoroughly surveyed the forward dynamic algorithms for serial rigid multibody systems and proposed an algorithm whose number of floating-point arithmetic operations grows linearly with the number of bodies of the system. Featherstone [52] has also developed fully recursive $O(n)$ forward dynamics algorithm for the open-chain systems using spatial vector notations. In spatial notation, the six-dimensional velocity and wrench vectors simplified the analysis of rigid body dynamics. By introducing the articulated body inertia, he was able to develop the $O(n)$ dynamics formulation for the forward dynamics where the joint accelerations are to be computed without explicit inversion of the mass matrix. Featherstone's $O(n)$ formulation has exerted a strong influence on the developments of tree-type[53], closed-loop systems [54], and manipulators [55]. A comparative study of these recursive methods is given by Stelzie et al. [56]. They concluded that the number of operations of recursive methods is influenced by the choice of the reference point, and the reference frame for the kinematic and dynamic equations. The number of operations is minimum for the case where the joint points are chosen as the local reference points, and the algorithm is evaluated in local reference frames of the bodies according to the Danvit-Hartenberg notations [57]. Recently, Saha [60] developed $O(n)$ algorithms for both the forward and inverse dynamics of serial multibody systems using decoupled natural orthogonal complement matrices. It is further extended for parallel manipulators in [61, 130, 131]. In a different perspective, Sohl and Borrow [62] developed the inverse dynamics algorithm based on Lie groups and Lie algebra.

1.2.3 Velocity transformation methods

The most efficient forward dynamics formulations of order $O(n)$ or $O(n^3)$, although they are theoretically different, have some common aspects where topology of the system plays a key role. Compared to the recursive formulations of open-loop systems, it is more complex to take into account the loop-closure constraints of closed-loop systems to obtain fully recursive $O(n)$ dynamic formulation [61, 64, 65]. Existing methodology for the dynamic modelling of closed-loop systems can be categorized as

- Cartesian coordinate approach, in which the system is represented using a set of coordinates for each body that specify the position of a point on that body and the set of parameters that specify the orientation of the body with respect to a fixed inertial frame.

- Minimal joint coordinate approach, in which the system is represented in terms of a set of minimal generalized coordinates.

The relative merits and demerits of the above two approaches are given in Table 1.1.

Table 1.1. Relative merits and demerits of dynamical formulations

Cartesian coordinate approach	Joint coordinate approach
Uses absolute Cartesian coordinates, i.e., 6 coordinates per body (a set of maximum generalized coordinates).	Uses relative joint coordinates. Reduced number of coordinates, hence, good numerical efficiency.
Coordinates are not based on joint and previous body, hence, independent of topology.	Position and orientation of a body depend on the configuration of its previous body and the DOF of the joint between them. Hence, it is a problem-dependent mathematical modeling.
Kinematics of each body is directly known.	Requires preprocessing and post processing to know the absolute motion of each body, which is arduous.
Formulation of dynamic equations is simple and straightforward. Easy to re-formulate if the architecture of the system is changed.	Derivation of dynamic equations is complex. Re-derivation of the equation of motion is required while architecture is changed.
Appropriate for implementation in general purpose multibody algorithms.	Appropriate for forward and inverse dynamic problems, where only generalized motion and applied forces are of interest.
Requires numerical methods to solve Differential Algebraic Equations (DAEs) which are less efficient and tend to be numerically unstable compared to the joint coordinate methods.	Requires numerical methods to solve Ordinary Differential Equations (ODE) which are efficient and numerically stable.
Constraint forces/moments are obtained as a part of the solution to DAEs.	Constraint forces/moments are eliminated. Therefore, it requires additional modeling effort to obtain them.
Relative coordinates are not readily available from the Cartesian coordinates which restrict modeling of feedback control that uses relative coordinates.	Relative coordinates are specially suited for open-chain systems and their control.

In order to take advantages of both the above methodologies, there exist several semi-recursive formulations for the closed-loop systems, where the equations of motion in the Cartesian coordinates are transformed into a set of the joint coordinates. Such formulations are based on the velocity transformation methods [66-77], where the advantages of both the formulations, i.e., the simplicity of the formulation in Cartesian coordinates and the efficiency of the formulation in joint coordinates, are kept. For open-loop systems, the relative joint coordinates are independent. As a result, the conversion of the large number of unconstrained NE equations to a reduced form of constrained equations using velocity transformation is straightforward. For the closed-loop systems, the relative joint coordinates are not independent because they have to satisfy the loop closure constraints. Usually the closed kinematic loops are cut to make them open and the Lagrange multipliers are introduced at the cut joints [67, 72]. The constrained equations of motion are then obtained for the resulting open system from the unconstrained NE equations, which are linear in the joint accelerations and the Lagrange multipliers. For a given motion of the system, once the Lagrange multipliers are known from the simultaneous solution of the constrained dynamic equations, the system can be treated as an open-loop system for which fully recursive algorithms can be developed.

Based on the review of the dynamic formalisms, the following points reveal:

1. Except the NE formulation, none of the dynamic formulations reveals direct information about the reaction forces and moments. The NE approach is simple, straightforward, and easy for general purpose algorithms.

2. The major drawback of the NE approach is that the dimension of the problem increases dramatically as the number of bodies increases, when compared to the joint space formulation.

3. Systematic reduction of the dimension of the unconstrained NE equations of motion can be done using velocity transformation methods.

4. One of the velocity transformation methods is to transform the unconstrained equations into the constrained ones using the decoupled natural orthogonal complement (DeNOC) matrices [60]. The method has emerged as a unified tool for recursive inverse and forward dynamics for serial [60] and parallel manipulators [61].

5. In the DeNOC formulation, two important physical quantities, namely, the twists and wrenches, are the velocity- and force-equivalents of the particle dynamics, respectively. The notion of twists and wrenches provides succinct and consistent expressions for rigid body dynamics.

6. Determination of the DeNOC matrices for a tree-type system is not available in the literature.

1.3 Balancing of Mechanisms

Related to the balancing of mechanisms the shaking force and shaking moment are defined as the resultant of the inertia forces and moments of moving links. When the dimensions of a mechanism and the input speed are given, the inertia forces depend only upon the mass distribution of the moving links. Balancing of shaking force and shaking moment in high speed mechanisms/machines reduces the forces transmitted to the frame. In effect, this reduces the noise and wear, improves the dynamic perfor- mance, and extends the fatigue life of the mechanisms. A considerable amount of research on balancing of shaking force and shaking moment in planar mechanisms has been carried out in the past, e.g. in [81-102]. In contrast, understanding of the balancing of spatial mechanisms is very li- mited [103-109]. This may be due to their complexity in formulation of the balancing problem.

To balance a mechanism completely, it is required to eliminate both the shaking force and the shaking moment. However, complete balancing of any one may result in an increased unbalance in the other one. Hence, the balancing problem of mechanisms can be postulated as an optimization problem [98-102, 115, 116]. Once the mathematical optimization problem is formulated, one can use any existing optimization tool to solve it. Since the dynamic performance characteristics, the shaking force, shaking mo- ment, input-torque, etc., depend on the mass and inertia of each link, and its mass center location [113], it is required to optimally distribute the link masses for dynamic balancing. A convenient way to represent the inertia properties of the links is treating them as the dynamically equivalent sys- tems of point masses. Such dynamically equivalent system is referred as *equimomental system* [110]. The concept of equimomental system of point-masses [110-112], and its effective utilization in mechanism ba- lancing have not been studied extensively, particularly, for the spatial mechanism.

2 Dynamics of Open-loop Systems

Mechanical systems are generally classified as open-loop (serial and/or tree type) and closed-loop systems according to their architecture. In this chapter, a dynamic formulation for open-loop systems is proposed to compute constraint wrenches, i.e., the reaction moments and forces at the joints. It is based on the Newton-Euler equations of motion and the concept of the DeNOC matrices. The DeNOC matrices are the logically split natural orthogonal complement (NOC) matrix that was originally proposed in [69]. Using the DeNOC matrices, Saha [60], Saha and Schiehlen [61], and Khan et al. [130, 131] developed a series of recursive inverse and forward dynamics algorithms for both serial and parallel kinematic chain systems. In those algorithms, joint reactions have been eliminated from the equations of motion. In this chapter, those reactions are of interest.

2.1 Kinematic Constraints in Serial Systems

One of the two types of open-loop systems, namely, serial-type, is considered in this section. Figure 2.1 shows serially connected $n+1$ rigid bodies labeled as #0 to #n—#0 being the base body.

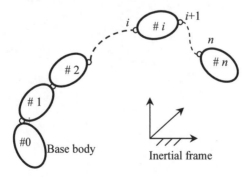

Fig. 2.1. Serial open-loop system

The bodies are coupled by n one-degree-of-freedom joints, revolute or prismatic, labeled as 1, ..., n. As shown in Fig. 2.1, the ith joint couples the $(i-1)$st body with the ith one. In order to concentrate on the mathematical formulation of the dynamic analysis, implementation details, for example, the selection of reference frames, representation of various vectors involved, transformation of vectors and matrices from one frame to other, etc. are not discussed here. They are explained in Appendix A.

Two definitions, namely, the *twist* and *wrench,* of a rigid body are, however, introduced which will be used throughout this book. For the angular velocity of a body, ω, and its linear velocity, $\mathbf{v}$, the *twist* is defined as the following 6-vector [68]:

$$\mathbf{t} \equiv \begin{bmatrix} \omega \\ \mathbf{v} \end{bmatrix} \tag{2.1}$$

Accordingly, if the moment and the force acting on the body denoted with $\mathbf{n}$ and $\mathbf{f}$, respectively, the 6-vector of *wrench* is defined as

$$\mathbf{w} \equiv \begin{bmatrix} \mathbf{n} \\ \mathbf{f} \end{bmatrix} \tag{2.2}$$

The above twist and wrench vectors are nothing but the spatial velocity and force of Featherstone [52] and Rodriguez et al. [55]. Their numerical values depend on both the reference point of a body with respect to which their linear velocity and the linear force are defined, and the reference frame in which they are represented. Considering ω_i and $\mathbf{v}_i$ as the angular velocity and the linear velocity of the origin point, O_i, on the ith body, Fig. 2.2, respectively, the twist, $\mathbf{t}_i$, is expressed in terms of its previous body, namely, the $(i-1)$st one, as [60]

$$\mathbf{t}_i = \mathbf{A}_{i,i-1}\mathbf{t}_{i-1} + \mathbf{p}_i \dot{\theta}_i \tag{2.3}$$

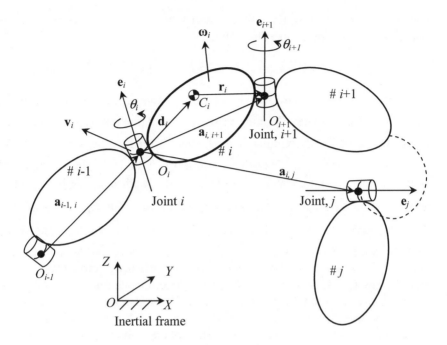

Fig. 2.2. Definitions of various vectors

where the 6×6 matrix, $\mathbf{A}_{i,i-1}$, is called the *twist propagation matrix* which transforms the twist of the $(i\text{-}1)$st body to the ith one as if they are rigidly connected. The matrix, $\mathbf{A}_{i,i-1}$, is given by

$$\mathbf{A}_{i,i-1} \equiv \begin{bmatrix} \mathbf{1} & \mathbf{O} \\ \tilde{\mathbf{a}}_{i,i-1} & \mathbf{1} \end{bmatrix} \tag{2.4}$$

in which the 3×3 skew-symmetric matrix, $\tilde{\mathbf{a}}_{i,i-1}$, associated with the 3-vector, $\mathbf{a}_{i,i-1}$, describes its cross product with an arbitrary 3-vector, $\mathbf{x}$, i.e.,

$$\tilde{\mathbf{a}}_{i,i-1}\mathbf{x} = \mathbf{a}_{i,i-1} \times \mathbf{x}, \text{ where } \tilde{\mathbf{a}}_{i,i-1} \equiv \begin{bmatrix} 0 & -a_{i,i-1z} & a_{i,i-1y} \\ a_{i,i-1z} & 0 & -a_{i,i-1x} \\ -a_{i,i-1y} & a_{i,i-1x} & 0 \end{bmatrix} \tag{2.5}$$

$a_{i,i-1x}$, $a_{i,i-1y}$ and $a_{i,i-1z}$ being the components of the vector, $\mathbf{a}_{i,i-1}$, from O_i to $O_{i\text{-}1}$, i.e., $\mathbf{a}_{i,i-1} = -\mathbf{a}_{i-1,1}$, as shown in Fig. 2.2. Moreover, for serially connected three bodies, namely, i, j, and k, the twist propagation matrices satisfy the following properties [60]:

$$\mathbf{A}_{ij}\mathbf{A}_{jk} = \mathbf{A}_{ik}\;;\; \mathbf{A}_{ii} = \mathbf{1}\;;\; \mathbf{A}_{ij}^{-1} = \mathbf{A}_{ji}\;;\; \text{and } \det(\mathbf{A}_{ij}) = 1 \qquad (2.6)$$

Now, the scalar, $\dot{\theta}_i$ of Eq. (2.3), is the joint rate, i.e., the time rate of the variable DH parameter, θ_i or b_i, as defined in Appendix A. For a revolute joint, the variable DH parameter is θ_i that defined as the angle of the ith body with respect to the $(i\text{-}1)$st one. For a prismatic joint, the variable DH parameter is b_i that defined displacement of the ith body with respect to the $(i\text{-}1)$st one. Hence, $\dot{b}_i$ is the linear speed of the point O_i with respect to $O_{i\text{-}1}$. Accordingly, $\dot{\theta}_i \equiv \dot{b}_i$ is to be used in Eq. (2.3). Note that O_i is the contact point between two links. For a revolute joint, point O_i is stationary on the $(i\text{-}1)$st and ith bodies, whereas for the prismatic joint, O_i, is stationary on the $(i\text{-}1)$st body but not on the ith one. Depending on the physical realization of a prismatic joint the converse can also be true, ie., O_i is stationary on the ith body but not on the $(i\text{-}1)$st one. For the ith body with a prismatic joint, O_i is the instant point of contact with the $(i\text{-}1)$st one. In Eq. (2.3), the 6-vector, $\mathbf{p}_i$, takes into account the motion of the ith body relative to the $(i\text{-}1)$st one, which is dependent on the degree of freedom (DOF) of the ith joint, and is called the *joint-rate propagation vector*. For one-DOF joints, e.g., revolute and prismatic,

$$\mathbf{p}_i \equiv \begin{bmatrix} \mathbf{e}_i \\ \mathbf{0} \end{bmatrix}: \text{Revolute joint; and } \mathbf{p}_i \equiv \begin{bmatrix} \mathbf{0} \\ \mathbf{e}_i \end{bmatrix}: \text{Prismatic joint} \qquad (2.7)$$

Other joints, e.g., spherical, cylindrical, etc., can be treated as the combination of revolute and prismatic joints [68]. In Eq.(2.7), the 3-vector, $\mathbf{e}_i$, represents the unit vector along the axis of rotation of the revolute joint, or along the direction of linear motion of the prismatic joint, respectively. Also, the matrix, $\mathbf{O}$, and the vector, $\mathbf{0}$, in Eqs. (2.4) and (2.7), respectively, are the 3×3 matrix and 3-vector of zeros, and the matrix, $\mathbf{1}$, is the 3×3 identity matrix. Henceforth, they will be understood as compatible sizes based on where they appear. Equation (2.3) establishes the recursive kinematical relation to determine the twist and twist rate of a body in the serial chain from its 0th body to the tip of the chain.

It is pointed out here that a rigid body undergoing a planar motion has three-DOF. Accordingly, the six-dimensional vectors are defined as three-dimensional vectors. For example, the 3-vectors of twist and wrench are as follows:

$$\mathbf{t}_i \equiv \begin{bmatrix} \omega_i \\ \mathbf{v}_i \end{bmatrix} ; \text{ and } \mathbf{w}_i \equiv \begin{bmatrix} n_i \\ \mathbf{f}_i \end{bmatrix} \tag{2.8}$$

where ω_i is the scalar angular velocity about the axis orthogonal to the plane of motion, and $\mathbf{v}_i$ is the 2-vector of linear velocity. Moreover, n_i is the angular moment about the normal to the plane of motion, and $\mathbf{f}_i$ is the 2-vector of force. Correspondingly, the 3×3 twist propagation matrix, $\mathbf{A}_{i,i-1}$, and the 3-vector of the joint-rate propagation vector, $\mathbf{p}_i$, are given by

$$\mathbf{A}_{i,i-1} \equiv \begin{bmatrix} 1 & \mathbf{0}^T \\ \overline{\mathbf{E}}\mathbf{a}_{i-1,i} & \mathbf{1} \end{bmatrix}; \tag{2.9a}$$

$$\mathbf{p}_i \equiv \begin{bmatrix} 1 \\ \mathbf{0} \end{bmatrix} \text{ for Revolute; } \mathbf{p}_i \equiv \begin{bmatrix} 0 \\ \mathbf{e}_i \end{bmatrix} \text{ for Prismatic}$$

In Eq. (2.9a), $\mathbf{0}$ and $\mathbf{e}_i$ are the 2-vectors of zeros and the unit vector along the motion of a prismatic joint, respectively, whereas $\mathbf{1}$ is the 2×2 identity matrix, and $\overline{\mathbf{E}}$ is given by

$$\overline{\mathbf{E}} \equiv \begin{bmatrix} 0 & -1 \\ 1 & 0 \end{bmatrix} \tag{2.9b}$$

Now, writing Eq. (2.3), for $i=1, \ldots, n$, and expressing them in a compact form, one can obtain the following:

$$\mathbf{A}_0 \mathbf{t}_0 + \mathbf{A}\mathbf{t} = \mathbf{N}_d \dot{\boldsymbol{\theta}} \tag{2.10}$$

where the $6n \times 6$, $6n \times 6n$, and $6n \times n$ matrices, $\mathbf{A}_0$, $\mathbf{A}$ and $\mathbf{N}_d$, are defined as:

$$\mathbf{A}_0 \equiv \begin{bmatrix} -\mathbf{A}_{10} \\ \mathbf{O} \\ \vdots \\ \mathbf{O} \end{bmatrix}; \mathbf{A} \equiv \begin{bmatrix} \mathbf{1} & \cdots & \mathbf{O} & \mathbf{O} \\ -\mathbf{A}_{21} & \cdots & \mathbf{O} & \mathbf{O} \\ \vdots & \ddots & \vdots & \vdots \\ \mathbf{O} & \cdots & -\mathbf{A}_{n,n-1} & \mathbf{1} \end{bmatrix}; \tag{2.11}$$

$$\mathbf{N}_d \equiv diag[\mathbf{p}_1 \quad \cdots \quad \mathbf{p}_n]$$

Also, the $6n$-vector of generalized twist, $\mathbf{t}$, and the n-vector of generalized joint rate, $\dot{\boldsymbol{\theta}}$, are defined as:

$$\mathbf{t} \equiv \begin{bmatrix} \mathbf{t}_1^T & \cdots & \mathbf{t}_n^T \end{bmatrix}^T \text{ and } \dot{\boldsymbol{\theta}} \equiv \begin{bmatrix} \dot{\theta}_1 & \cdots & \dot{\theta}_n \end{bmatrix}^T \tag{2.12}$$

The generalized twist, $\mathbf{t}$, is then obtained from Eq. (2.10) by inverting the matrix, $\mathbf{A}$, as

$$\mathbf{t} = -\mathbf{N}_l\mathbf{A}_0\mathbf{t}_0 + \mathbf{N}_l\mathbf{N}_d\dot{\boldsymbol{\theta}} \tag{2.13}$$

where

$$\mathbf{N}_l \equiv \mathbf{A}^{-1} = \begin{bmatrix} \mathbf{1} & \mathbf{O} & \cdots & \mathbf{O} \\ \mathbf{A}_{21} & \mathbf{1} & \vdots & \mathbf{O} \\ \vdots & \vdots & \ddots & \vdots \\ \mathbf{A}_{n1} & \mathbf{A}_{n2} & \cdots & \mathbf{1} \end{bmatrix} \tag{2.14}$$

In Eq. (2.14), $\mathbf{A}_{i,j} = \mathbf{A}_{i,i-1} \cdots \mathbf{A}_{j+1,j}$ for $i > j$. Note that the inverse of the $6n \times 6n$ matrix, $\mathbf{A}$, from Eq. (2.11), may appear to be difficult and computationally expensive. But it is not true. It can be easily obtained from the fact that $\mathbf{N}_l\mathbf{A} = \mathbf{1}$.

For the serial open-loop system, Fig. 2.1, if the base body is fixed, i.e., $\mathbf{t}_0 = \mathbf{0}$, Eq. (2.13) takes the following simple form:

$$\mathbf{t} = \mathbf{N}\dot{\theta}, \text{ where } \mathbf{N} \equiv \mathbf{N}_l\mathbf{N}_d \tag{2.15}$$

The $6n \times 6n$ lower block triangular matrix, $\mathbf{N}_l$, and the $6n \times n$ block diagonal matrix, $\mathbf{N}_d$, are called the decoupled natural orthogonal complement (DeNOC) matrices, as derived in [60], whereas $\mathbf{N}$ is the natural orthogonal complement (NOC) matrix of [68]. The word 'natural' is used in [60, 68] because the matrix, $\mathbf{N}$, is derived naturally from the velocity constraints, Eq. (2.3), without resorting to any numerical methods, e.g., in [32] and others. The DeNOC matrices, $\mathbf{N}_l$ and $\mathbf{N}_d$, separate the architecture information of the serial bodies from the joint information. For example, if the ith joint is locked, i.e., $\dot{\theta}_i$ is zero, the matrix, $\mathbf{A}_{i,i-1}$, transfers the twist of #(i-1) to #i, as if they have formed a rigid composite body.

2.2 Kinematic Constraints in Tree-type Systems

A tree-type open-loop system is shown in Fig. 2.3. The DeNOC matrices for the tree-type system are derived in this section, which forms the basis for the dynamic modeling of closed-loop systems presented in Chapter 3. All serial chain branches of the tree are identified by their respective base bodies. Let us define the longest chain as the main chain connected to the fixed body or the floating base, numbered as $\#0^0$, and has n^0 bodies. All other subchains called branches are assumed to be connected to the main chain. A subchain k with n^k bodies is defined as the one that stems from the kth body of the main chain, i.e., $\#k^0$. Similarly, subchain ℓ is connected to the ℓth body of the main chain and has n^ℓ bodies, as shown in Fig. 2.3. Moreover, if the joints in the system are either revolute and/or prismatic, and the base is fixed, the DOF of the system is equal to the number of joints or moving bodies. Without any loss of generality, the DeNOC matrices are determined by assuming all joints are either revolute or prismatic.

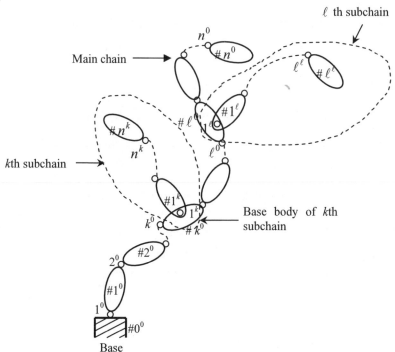

Fig. 2.3. A tree-type system

From Eq. (2.13), if the base body is fixed, i.e., $\mathbf{t}_0 = \mathbf{0}$, the generalized twist for the main chain, denoted as $\mathbf{t}^0$, is obtained as

$$\mathbf{t}^0 = \mathbf{N}_I^0 \mathbf{N}_d^0 \dot{\boldsymbol{\theta}}^0 \tag{2.16}$$

where superscript '0' indicates the main chain. Similar to the derivation of Eq. (2.13), the generalized twist of the subchain k, whose base body is $\#0^k$, or $\#k^0$ of the main chain, denoted with the $6n^k$-vector, $\mathbf{t}^k$, is given by

$$\mathbf{t}^k = -\mathbf{N}_I^k \mathbf{A}_0^k \mathbf{t}_0^k + \mathbf{N}_I^k \mathbf{N}_d^k \dot{\boldsymbol{\theta}}^k \tag{2.17}$$

In which the $6n^k \times 6n^k$ matrix, $\mathbf{N}_I^k$, and the $6n^k \times n^k$ matrix, $\mathbf{N}_d^k$, are the DeNOC matrices of the kth subchain, and $\mathbf{t}_0^k \equiv \mathbf{t}_k^0$. The twist, $\mathbf{t}_k^0$, is obtained from Eq (2.16) as

$$\mathbf{t}_k^0 = \mathbf{N}_{lk}^0 \mathbf{N}_d^0 \dot{\boldsymbol{\theta}}^0 \tag{2.18}$$

In Eq. (2.18), the $6 \times 6n^0$ matrix, $\mathbf{N}_{lk}^0$, is given by

$$\mathbf{N}_{lk}^0 \equiv \begin{bmatrix} \mathbf{A}_{k,1}^0 & \cdots & \mathbf{A}_{k,k-1}^0 & \mathbf{1} & \mathbf{O} & \cdots & \mathbf{O} \end{bmatrix} \tag{2.19}$$

Substituting Eq. (2.18) into Eq. (2.17) yields

$$\mathbf{t}^k = \mathbf{N}_I^{k0} \mathbf{N}_d^0 \dot{\boldsymbol{\theta}}^0 + \mathbf{N}_I^k \mathbf{N}_d^k \dot{\boldsymbol{\theta}}^k, \text{ where } \mathbf{N}_I^{k0} \equiv -\mathbf{N}_I^k \mathbf{A}_0^k \mathbf{N}_{lk}^0 \tag{2.20}$$

Similarly, for another subchain, ℓ, one can also write the $6n^\ell$-vector of generalized twist, $\mathbf{t}^\ell$, from Eq. (2.20) as

$$\mathbf{t}^\ell = \mathbf{N}_I^{\ell 0} \mathbf{N}_d^0 \dot{\boldsymbol{\theta}}^0 + \mathbf{N}_I^\ell \mathbf{N}_d^\ell \dot{\boldsymbol{\theta}}^\ell, \text{ where } \mathbf{N}_I^{\ell 0} \equiv -\mathbf{N}_I^\ell \mathbf{A}_0^\ell \mathbf{N}_{l\ell}^0 \tag{2.21}$$

where the $6n^\ell \times 6$, $6n^\ell \times n^\ell$, $6n^\ell \times 6n^\ell$, and $6 \times 6n^\ell$ matrices $\mathbf{A}_0^\ell$, $\mathbf{N}_d^\ell$, $\mathbf{N}_I^\ell$, and $\mathbf{N}_{l\ell}^0$, respectively, are defined similar to those for the subchain k. Combining Eqs. (2.16), (2.20) and (2.21), the generalized twist, $\mathbf{t}$, for the tree-type system shown in Fig. 2.3 with base body, $\#0^0$, fixed, i.e., $\mathbf{t}_0^0 = \mathbf{0}$, is given by

$$\mathbf{t} = \mathbf{N}\dot{\boldsymbol{\theta}}, \text{ where } \mathbf{N} \equiv \mathbf{N}_I \mathbf{N}_d \tag{2.22}$$

and the $6n$-vector—$n \equiv n^0 + n^k + n^\ell$ being the total number of moving bodies in the tree-type system— $\mathbf{t}$, the n-vector, $\dot{\boldsymbol{\theta}}$, the $6n \times 6n$ matrix, $\mathbf{N}_I$, and the $6n \times n$ matrix, $\mathbf{N}_d$, are defined as:

$$\mathbf{t} \equiv \begin{bmatrix} \mathbf{t}^0 \\ \mathbf{t}^k \\ \mathbf{t}^\ell \end{bmatrix} ; \ \dot{\boldsymbol{\theta}} \equiv \begin{bmatrix} \dot{\boldsymbol{\theta}}^0 \\ \dot{\boldsymbol{\theta}}^k \\ \dot{\boldsymbol{\theta}}^\ell \end{bmatrix} ; \ \mathbf{N}_l \equiv \begin{bmatrix} \mathbf{N}_l^0 & \mathbf{O} & \mathbf{O} \\ \mathbf{N}_l^{k0} & \mathbf{N}_l^k & \mathbf{O} \\ \mathbf{N}_l^{\ell 0} & \mathbf{O} & \mathbf{N}_l^\ell \end{bmatrix} ; \text{and} \tag{2.23}$$

$$\mathbf{N}_d \equiv \begin{bmatrix} \mathbf{N}_d^0 & \mathbf{O} & \mathbf{O} \\ \mathbf{O} & \mathbf{N}_d^k & \mathbf{O} \\ \mathbf{O} & \mathbf{O} & \mathbf{N}_d^\ell \end{bmatrix}$$

For additional subchains, one can modify the expressions of $\mathbf{t}$, $\dot{\boldsymbol{\theta}}$, $\mathbf{N}_l$, and $\mathbf{N}_d$, as given by Eq. (2.23). Equations (2.22) and (2.23) provide the De-NOC matrices for the tree-type system, which will be used to reduce the dimension of the system's NE equations of motion.

2.3 Equations of Motion

The Newton-Euler (NE) equations of motion for the ith rigid body of a multibody system are written from its free-body diagram as [4, 10]

$$\mathbf{I}_i^c \dot{\boldsymbol{\omega}}_i + \tilde{\boldsymbol{\omega}}_i \mathbf{I}_i^c \boldsymbol{\omega}_i = \mathbf{n}_i^c \tag{2.24}$$

$$m_i \dot{\mathbf{v}}_i^c = \mathbf{f}_i^c \tag{2.25}$$

where $\mathbf{n}_i^c$ is the resultant of all the external moments about its mass center, C_i, and $\mathbf{f}_i^c$ is the resultant force acting at C_i. Moreover, $\mathbf{I}_i^c$ is the inertia tensor with respect to C_i. To express the NE equations of the ith body in terms of its twist and wrench defined with respect to the origin, O_i, the velocity, $\mathbf{v}_i$, and acceleration, $\dot{\mathbf{v}}_i$, are to be expressed as

$$\mathbf{v}_i = \mathbf{v}_i^c + \tilde{\mathbf{d}}_i \boldsymbol{\omega}_i \tag{2.26}$$

$$\dot{\mathbf{v}}_i = \dot{\mathbf{v}}_i^c + \tilde{\mathbf{d}}_i \dot{\boldsymbol{\omega}}_i + \tilde{\boldsymbol{\omega}}_i \tilde{\mathbf{d}}_i \boldsymbol{\omega}_i \tag{2.27}$$

where $\tilde{\mathbf{d}}_i$ and $\tilde{\boldsymbol{\omega}}_i$ are the 3×3 skew-symmetric matrices associated with the 3-vectors, $\mathbf{d}_i$ and $\boldsymbol{\omega}_i$, respectively. Consequently, the moment and the force, $\mathbf{n}_i$ and $\mathbf{f}_i$, respectively, with respect to O_i are obtained from those with respect to C_i as:

$$\mathbf{n}_i = \mathbf{n}_i^c + \mathbf{d}_i \times \mathbf{f}_i^c \text{ and } \mathbf{f}_i = \mathbf{f}_i^c \tag{2.28}$$

On substitution of Eqs. (2.27) and (2.28) into Eqs. (2.24) and (2.25), one obtains the following:

$$\mathbf{I}_i \dot{\boldsymbol{\omega}}_i + m_i \tilde{\mathbf{d}}_i \dot{\mathbf{v}}_i + \tilde{\boldsymbol{\omega}}_i \mathbf{I}_i \boldsymbol{\omega}_i = \mathbf{n}_i \tag{2.29}$$

$$m_i \dot{\mathbf{v}}_i - m_i \tilde{\mathbf{d}}_i \dot{\boldsymbol{\omega}}_i - m_i \tilde{\boldsymbol{\omega}}_i \tilde{\mathbf{d}}_i \boldsymbol{\omega}_i = \mathbf{f}_i \tag{2.30}$$

Note from the parallel axis theorem, $\mathbf{I}_i = \mathbf{I}_i^c - m_i \tilde{\mathbf{d}}_i^2$, which is the inertia matrix about O_i. Equations (2.29) and (2.30) are now written in a compact form as [61]:

$$\mathbf{M}_i \dot{\mathbf{t}}_i + \mathbf{W}_i \mathbf{M}_i \mathbf{E}_i \mathbf{t}_i = \mathbf{w}_i \tag{2.31}$$

where the 6×6 matrices of the mass, $\mathbf{M}_i$, and of the angular velocity , $\mathbf{W}_i$, and the 6-vector of external wrench, $\mathbf{w}_i$, are defined as

$$\mathbf{M}_i \equiv \begin{bmatrix} \mathbf{I}_i & m_i \tilde{\mathbf{d}}_i \\ -m_i \tilde{\mathbf{d}}_i & m_i \mathbf{1} \end{bmatrix}; \ \mathbf{W}_i \equiv \begin{bmatrix} \tilde{\boldsymbol{\omega}}_i & \mathbf{O} \\ \mathbf{O} & \tilde{\boldsymbol{\omega}}_i \end{bmatrix}; \text{ and } \mathbf{w}_i \equiv \begin{bmatrix} \mathbf{n}_i \\ \mathbf{f}_i \end{bmatrix} \tag{2.32}$$

In Eq. (2.32), $\mathbf{M}_i$ is the mass matrix that embodies the mass and inertia properties of the ith body about O_i , whereas the 6×6 coupling matrix, $\mathbf{E}_i$ of Eq.(2.31), is given by

$$\mathbf{E}_i \equiv \begin{bmatrix} \mathbf{1} & \mathbf{O} \\ \mathbf{O} & \mathbf{O} \end{bmatrix} \tag{2.33}$$

Writing Eq. (2.31) for a system of n moving bodies, i.e., for $i=1, \ldots, n$, the $6n$ scalar equations of motion are expressed as

$$\mathbf{M}\dot{\mathbf{t}} + \mathbf{W}\mathbf{M}\mathbf{E}\mathbf{t} = \mathbf{w} \tag{2.34}$$

where the $6n \times 6n$ matrices, $\mathbf{M}$, $\mathbf{W}$, and $\mathbf{E}$, are the generalized mass, angular velocity, and coupling matrices, respectively, namely,

$$\mathbf{M} \equiv diag[\mathbf{M}_1 \ \cdots \ \mathbf{M}_n]; \ \mathbf{W} \equiv diag[\mathbf{W}_1 \ \cdots \ \mathbf{W}_n]; \tag{2.35}$$
$$\mathbf{E} \equiv diag[\mathbf{E}_1 \ \cdots \ \mathbf{E}_n]$$

Also, the $6n$-vectors of the generalized twist, twist-rate, and wrench, $\mathbf{t}$, $\dot{\mathbf{t}}$, and $\mathbf{w}$, respectively, are

$$\mathbf{t} \equiv [\mathbf{t}_1^T \ \cdots \ \mathbf{t}_n^T]^T, \ \dot{\mathbf{t}} \equiv [\dot{\mathbf{t}}_1^T \ \cdots \ \dot{\mathbf{t}}_n^T]^T, \text{ and } \mathbf{w} \equiv [\mathbf{w}_1^T \ \cdots \ \mathbf{w}_n^T]^T \tag{2.36}$$

The counterpart of Eqs. (2.31)-(2.36) for planar motion are as follows:

$$\mathbf{M}_i\dot{\mathbf{t}}_i + \mathbf{C}_i\mathbf{t}_i = \mathbf{w}_i \qquad (2.37)$$

where $\mathbf{t}_i, \dot{\mathbf{t}}_i$ and $\mathbf{w}_i$, are the 3-vectors, i.e.,

$$\mathbf{t}_i \equiv \begin{bmatrix} \omega_i \\ \mathbf{v}_i \end{bmatrix}, \ \dot{\mathbf{t}}_i \equiv \begin{bmatrix} \dot{\omega}_i \\ \dot{\mathbf{v}}_i \end{bmatrix}, \text{ and } \mathbf{w}_i \equiv \begin{bmatrix} n_i \\ \mathbf{f}_i \end{bmatrix} \qquad (2.38)$$

In Eq. (2.38), ω_i $\dot{\omega}_i$ and n_i are scalars, and $\mathbf{v}_i, \dot{\mathbf{v}}_i$, and $\mathbf{f}_i$ are the 2-vectors. The 3×3 matrices, $\mathbf{M}_i$ and $\mathbf{C}_i$ are now defined for planar motion as:

$$\mathbf{M}_i \equiv \begin{bmatrix} I_i & -m_i\mathbf{d}_i^T\overline{\mathbf{E}} \\ m_i\overline{\mathbf{E}}\mathbf{d}_i & m_i\mathbf{1} \end{bmatrix} \text{ and } \mathbf{C}_i \equiv \begin{bmatrix} 0 & \mathbf{0}^T \\ -m_i\omega_i\mathbf{d}_i & \mathbf{O} \end{bmatrix} \qquad (2.39)$$

In which $\mathbf{1}$ and $\mathbf{O}$ are the 2×2 identity and zero matrices, respectively, and $\mathbf{0}$ is the 2-vector of zeros, whereas the 2×2 matrix, $\overline{\mathbf{E}}$, is defined in Eq. (2.9b). Combining the equations of motion, Eq. (2.37), for $i=1, \ldots, n$, for the n moving bodies of a mechanism, one gets

$$\mathbf{M}\dot{\mathbf{t}} + \mathbf{C}\mathbf{t} = \mathbf{w} \qquad (2.40)$$

where the $3n×3n$ matrices are $\mathbf{M} \equiv diag[\mathbf{M}_1 \ \cdots \ \mathbf{M}_n]$ and $\mathbf{C} \equiv diag[\mathbf{C}_1 \ \cdots \ \mathbf{C}_n]$. Also, the $3n$-vectors of generalized twist, twist-rate and wrench vectors, $\mathbf{t}, \dot{\mathbf{t}}$ and $\mathbf{w}$, respectively, are defined in accordance with Eq. (2.36).

The expression in the left hand side of Eq. (2.34) or (2.40) denotes the effective inertia forces and moments, and that on the right hand side represents the external forces and moments, and those due to the constraints at the joints. The $6n$ and $3n$ scalar equations of motion, Eq. (2.34) and (2.40), are the unconstrained equations of motion of a multibody system moving in space and plane, respectively. Using these equations, if one is interested to know all the joint forces and toques that include the reactions, as well as the driving forces and torques for a given set of body motions, all the $6n$ or $3n$ scalar equations have to be solved simultaneously. For example, for a planar four-bar linkage, one needs to solve nine equations to find all the unknown reactions and the driving torque. Such determination of the joint forces and torques requires order (n^3) calculations, which are not computationally efficient, particularly, for large n. Moreover, in reaction or constraint force optimization, these calculations are to be repeated hundreds of time. Hence, an efficient algorithm is sought to solve the constraint forces, as explained next.

2.4 Constraint Wrench for Serial Systems

Referring to Fig. 2.4, the dynamic equations of motion of the ith body with respect to O_i, Eqs. (2.29) and (2.30), are re-written in terms of the reactions at the neighboring joints as

$$\mathbf{n}_{i-1,i} - \mathbf{n}_{i,i+1} - \mathbf{a}_{i,i+1} \times \mathbf{f}_{i,i+1} = \mathbf{n}_i^* - \mathbf{n}_i^e \qquad (2.41)$$

$$\mathbf{f}_{i-1,i} - \mathbf{f}_{i,i+1} = \mathbf{f}_i^* - \mathbf{f}_i^e \qquad (2.42)$$

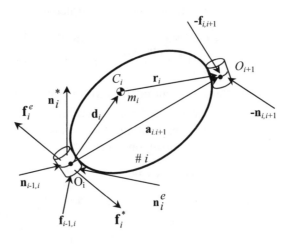

Fig. 2.4. Free body diagram of the ith body

where $\mathbf{n}_i^*$ and $\mathbf{f}_i^*$, respectively, are the resultant inertia moment and force due to the motion of the ith body, as given by the left hand sides of Eqs. (2.29) and (2.30), i.e., $\mathbf{n}_i^* \equiv \mathbf{I}_i \dot{\boldsymbol{\omega}}_i + m_i \tilde{\mathbf{d}}_i \dot{\mathbf{v}}_i + \tilde{\boldsymbol{\omega}}_i \mathbf{I}_i \boldsymbol{\omega}_i$ and $\mathbf{f}_i^* \equiv m_i \dot{\mathbf{v}}_i - m_i \tilde{\mathbf{d}}_i \dot{\boldsymbol{\omega}}_i - m_i \tilde{\boldsymbol{\omega}}_i \tilde{\mathbf{d}}_i \boldsymbol{\omega}_i$, whereas the right hand sides of Eqs. (2.29) and (2.30) are, $\mathbf{n}_i \equiv \mathbf{n}_i^e + \mathbf{n}_{i-1,i} - \mathbf{n}_{i,i+1} - \mathbf{a}_{i,i+1} \times \mathbf{f}_{i,i+1}$ and $\mathbf{f}_i \equiv \mathbf{f}_i^e + \mathbf{f}_{i-1,i} - \mathbf{f}_{i,i+1}$. Moreover, the moment, $\mathbf{n}_{i-1,i}$, and the force, $\mathbf{f}_{i-1,i}$, are those applied by the $(i-1)$st body to the ith one at the ith joint. Similarly, $\mathbf{n}_{i,i+1}$ and $\mathbf{f}_{i,i+1}$ are those from the ith body to the $(i+1)$st one at the $(i+1)$st joint. Furthermore, the external moment, $\mathbf{n}_i^e$, and force, $\mathbf{f}_i^e$, are those due to gravity, dissipation, environment, etc., and considered with respect to O_i, as shown in Fig. 2.4. Equations (2.41) and (2.42) are now put in a compact form as

$$\mathbf{w}_{i-1,i} = \mathbf{A}'_{i,i+1}\mathbf{w}_{i,i+1} + \mathbf{w}_i^* - \mathbf{w}_i^e \qquad (2.43)$$

where the 6-vectors, $\mathbf{w}_{i-1,i}$, $\mathbf{w}_i^*$ and $\mathbf{w}_i^e$, are defined as the *constraint joint, inertia,* and *external wrenches,* respectively, i.e.,

$$\mathbf{w}_{i-1,i} \equiv \begin{bmatrix} \mathbf{n}_{i-1,i} \\ \mathbf{f}_{i-1,i} \end{bmatrix}; \ \mathbf{w}_i^* \equiv \begin{bmatrix} \mathbf{n}_i^* \\ \mathbf{f}_i^* \end{bmatrix}; \text{ and } \mathbf{w}_i^e \equiv \begin{bmatrix} \mathbf{n}_i^e \\ \mathbf{f}_i^e \end{bmatrix} \qquad (2.44)$$

In Eq. (2.44), the 6×6 matrix, $\mathbf{A}'_{i,i+1}$, is the *wrench propagation matrix* which transforms the wrench acting at point O_{i+1} to O_i of the ith body, i.e.,

$$\mathbf{A}'_{i,i+1} \equiv \begin{bmatrix} \mathbf{1} & \widetilde{\mathbf{a}}_{i,i+1} \\ \mathbf{O} & \mathbf{1} \end{bmatrix} \qquad (2.45)$$

where $\widetilde{\mathbf{a}}_{i,i+1}$ is the 3×3 skew-symmetric matrix associated with the 3-vector, $\mathbf{a}_{i,i+1}$, as shown in Fig. 2.4. Similar to the properties of twist propagation matrix, Eq. (2.6), wrench propagation matrix also has the following properties:

$$\mathbf{A}'_{i-1,i}\mathbf{A}'_{i,i+1} = \mathbf{A}'_{i-1,i+1}, \ \mathbf{A}'_{ii} = \mathbf{1}, \text{ and } \mathbf{A}'^{-1}_{i,i+1} = \mathbf{A}'_{i+1,i} \qquad (2.46)$$

Note that in Eq. (2.46), $\mathbf{A}'_{i,i+1} = \mathbf{A}^T_{i+1,i}$, where $\mathbf{A}_{i+1,i}$ is the twist-propagation matrix defined similar to Eq. (2.4). The twist propagation matrices transform the twists outwardly, i.e., from the base body to the tip body of a serial system, whereas the wrench propagation matrices transform the wrenches inwardly, i.e., from the tip body to the base body. This shows the duality of the motion and corresponding force. Hence, the projection of the constraint joint wrench onto the joint axis gives the driving torque/force, τ_i, i.e.,

$$\tau_i = \mathbf{p}_i^T \mathbf{w}_{i-1,i} \qquad (2.47)$$

Equation (2.47) is the famous NE recursive formulation for the inverse dynamics of a serial robotic system [60]. However, the interest here is to compute the constraint wrenches, $\mathbf{w}_{i-1,i}$ and $\mathbf{w}_{i,i+1}$, recursively using Eq. (2.43) for given external and inertia wrenches, $\mathbf{w}_i^e$ and $\mathbf{w}_i^*$, respectively. These wrenches are essential for the mechanical design of the links, bearings, and also for the constraint force optimization. Such aspects are generally neglected in the literature. Equation (2.43) is now written, for $i=1, \ldots, n$, and put in a compact form as

$$
\begin{bmatrix}
1 & -\mathbf{A}'_{12} & \cdots & \mathbf{O} \\
\vdots & \vdots & \cdots & \mathbf{O} \\
\mathbf{O} & \mathbf{O} & \ddots & -\mathbf{A}'_{n-1,n} \\
\mathbf{O} & \mathbf{O} & \cdots & 1
\end{bmatrix}
\begin{bmatrix}
\mathbf{w}_{01} \\
\vdots \\
\mathbf{w}_{n-2,n-1} \\
\mathbf{w}_{n-1,n}
\end{bmatrix}
=
\begin{bmatrix}
\mathbf{w}'_1 \\
\vdots \\
\mathbf{w}'_{n-1} \\
\mathbf{w}'_n
\end{bmatrix}
\tag{2.48}
$$

where, $\mathbf{w}'_i \equiv \mathbf{w}^*_i - \mathbf{w}^e_i$. Denoting the $6n$-vectors, $\tilde{\mathbf{w}} \equiv \begin{bmatrix} \mathbf{w}^T_{01} & \mathbf{w}^T_{12} & \cdots & \mathbf{w}^T_{n-1,n} \end{bmatrix}^T$, and $\mathbf{w}' \equiv \begin{bmatrix} \mathbf{w}'^T_1 & \mathbf{w}'^T_2 & \cdots & \mathbf{w}'^T_n \end{bmatrix}^T$, vector $\tilde{\mathbf{w}}$ is obtained from Eq. (2.48) as

$$\tilde{\mathbf{w}} = \mathbf{N}_u \mathbf{w}' \tag{2.49}$$

In which the $6n \times 6n$ matrix, $\mathbf{N}_u$, is the inverse of the coefficient matrix of $\tilde{\mathbf{w}}$ appearing in Eq. (2.48) and obtained similar to Eq. (2.14) as

$$
\mathbf{N}_u \equiv
\begin{bmatrix}
1 & \mathbf{A}'_{12} & \cdots & \mathbf{A}'_{1n} \\
\mathbf{O} & 1 & \cdots & \mathbf{A}'_{2n} \\
\vdots & \vdots & \ddots & \vdots \\
\mathbf{O} & \mathbf{O} & \mathbf{O} & 1
\end{bmatrix}
\tag{2.50}
$$

where $\mathbf{A}'_{i,j} = \mathbf{A}'_{i,i-1} \cdots \mathbf{A}'_{j+1,j}$ for $i<j$. The $6n \times 6n$ upper block triangular matrix, $\mathbf{N}_u$, is nothing but the transpose of one of the DeNOC matrices, $\mathbf{N}_l$ of Eq. (2.14), i.e,

$$\mathbf{N}_u = \mathbf{N}^T_l \tag{2.52}$$

Upon pre-multiplication of the transpose of $\mathbf{N}_d$, the block diagonal matrix of the DeNOC matrices, to both sides of Eq. (2.49) yields

$$\mathbf{N}^T_d \tilde{\mathbf{w}} = \mathbf{N}^T_d \mathbf{N}_u \mathbf{w}' = \mathbf{N}^T_d \mathbf{N}^T_l \mathbf{w}' = \mathbf{N}^T \mathbf{w}' \tag{2.53}$$

where Eq. (2.52) is used. Moreover, the left hand of Eq. (2.53) is nothing but $\boldsymbol{\tau} \equiv \begin{bmatrix} \tau_1 & \cdots & \tau_n \end{bmatrix}^T$ due to Eq. (2.47). Hence, Eq. (2.53) is rewritten as

$$\boldsymbol{\tau} = \mathbf{N}^T \mathbf{w}' \tag{2.54}$$

where $\boldsymbol{\tau}$ is the n-vector of generalized driving torques/forces. Equations (2.49) and (2.54) are the high-level description of the constraint wrenches of the system under study and the driving torques/forces obtained in the recursive NE inverse dynamics algorithm from their associated algorithmic descriptions given by Eqs. (2.43) and (2.47), respectively.

2.5 Constraint Wrench in Tree-type Systems

Referring to Fig. 2.3, the constraint wrench of the ith body belonging to the main chain, 0, is written from Eq. (2.43) as

$$\mathbf{w}_{i-1,i}^0 = \mathbf{A}_{i,i+1}'^0 \mathbf{w}_{i,i+1}^0 + \mathbf{w}_i^{*0} - \mathbf{w}_i^{e0} \qquad\qquad for \ i \neq k, \ell \qquad (2.55a)$$

$$\mathbf{w}_{i-1,i}^0 = \mathbf{A}_{i,i+1}'^0 \mathbf{w}_{i,i+1}^0 + \mathbf{A}_{0,1}'^i \mathbf{w}_{0,1}^i + \mathbf{w}_i^{*0} - \mathbf{w}_i^{e0} \qquad for \ i = k, \ell \qquad (2.55b)$$

Note in Eq. (2.55b) that the effect of subchain k and ℓ on the kth and ℓth body of the mainchain, respectively, is taken into account through the second term, namely, $\mathbf{A}_{0,1}'^i \mathbf{w}_{0,1}^i$. Now, for each subchain, whose bodies are serially connected, the constraint wrench formulation is exactly same as given in Section 2.4. However, the superscript index is used to denote a subchain. Combining Eqs. (2.55a) and (2.55.b) for all bodies of the main chain, one gets

$$\widetilde{\mathbf{w}}^0 = \mathbf{N}_u^0 \mathbf{w}'^0 + \mathbf{N}_u^0 \mathbf{A}_0'^k \mathbf{w}_{01}^k + \mathbf{N}_u^0 \mathbf{A}_0'^\ell \mathbf{w}_{01}^\ell , \qquad (2.56)$$

$$\text{where } \mathbf{A}_0'^k \equiv \begin{bmatrix} \mathbf{O} \\ \vdots \\ \mathbf{A}_{01}'^k \\ \vdots \\ \mathbf{O} \end{bmatrix} \text{ and } \mathbf{A}_0'^\ell \equiv \begin{bmatrix} \mathbf{O} \\ \vdots \\ \mathbf{A}_{01}'^\ell \\ \vdots \\ \mathbf{O} \end{bmatrix}$$

and the 6×6 matrices, $\mathbf{A}_0'^k$ and $\mathbf{A}_0'^\ell$, occupy the block rows corresponding to the kth and ℓth bodies of the main chain, respectively. In addition, for the subchains, k and ℓ, one obtains the following from Eq. (2.49)

$$\widetilde{\mathbf{w}}^k = \mathbf{N}_u^k \mathbf{w}'^k \qquad \text{and} \quad \widetilde{\mathbf{w}}^\ell = \mathbf{N}_u^\ell \mathbf{w}'^\ell \qquad (2.57)$$

Using Eq. (2.57) and the definitions after Eq. (2.48), the 6-vector, $\mathbf{w}_{01}$, for the subchains , k and ℓ, namely, $\mathbf{w}_{01}^k$ and $\mathbf{w}_{01}^\ell$, respectively, are obtained as

$$\mathbf{w}_{01}^{k} = \mathbf{N}_{uk}^{k}\mathbf{w}'^{k} \text{ and } \mathbf{w}_{01}^{\ell} = \mathbf{N}_{u\ell}^{\ell}\mathbf{w}'^{\ell} \tag{2.58}$$

where the $6\times6n^{k}$ and $6\times6n^{\ell}$ matrices, $\mathbf{N}_{uk}^{k}$ and $\mathbf{N}_{u\ell}^{\ell}$, respectively, are defined as $\mathbf{N}_{uk}^{k} = \begin{bmatrix} \mathbf{1} & \mathbf{A}_{12}'^{k} & \cdots & \mathbf{A}_{1n^{k}}'^{k} \end{bmatrix}$ and $\mathbf{N}_{u\ell}^{\ell} = \begin{bmatrix} \mathbf{1} & \mathbf{A}_{12}'^{\ell} & \cdots & \mathbf{A}_{1n^{\ell}}'^{\ell} \end{bmatrix}$. Upon substitution of Eq.(2.58) into Eq. (2.56) then yields

$$\widetilde{\mathbf{w}} = \mathbf{N}_{u}^{0}\mathbf{w}'^{0} + \mathbf{N}_{u}^{0k}\mathbf{w}'^{k} + \mathbf{N}_{u}^{0\ell}\mathbf{w}'^{\ell} \tag{2.59}$$

where $\mathbf{N}_{u}^{0k} \equiv \mathbf{N}_{u}^{0}\mathbf{A}_{01}'^{k}\mathbf{N}_{uk}^{k}$ and $\mathbf{N}_{u}^{0\ell} \equiv \mathbf{N}_{u}^{0}\mathbf{A}_{01}'^{\ell}\mathbf{N}_{u\ell}^{\ell}$. Next, Eqs. (2.57) and (2.59) are combined to obtain the generalized constraint joint wrench for the tree-type system shown in Fig. 2.3, i.e., $\widetilde{\mathbf{w}}$, in terms of its generalized body wrench, $\mathbf{w}'$, i.e.,

$$\widetilde{\mathbf{w}} = \mathbf{N}_{u}\mathbf{w}' \tag{2.60}$$

where the $6n$-vectors , $\widetilde{\mathbf{w}}$ and $\mathbf{w}'$, and the $6n\times6n$ upper block triangular matrix $\mathbf{N}_{u}$ are defined as:

$$\widetilde{\mathbf{w}} \equiv \begin{bmatrix} \widetilde{\mathbf{w}}^{0} \\ \widetilde{\mathbf{w}}^{k} \\ \widetilde{\mathbf{w}}^{\ell} \end{bmatrix}; \mathbf{w}' \equiv \begin{bmatrix} \mathbf{w}'^{0} \\ \mathbf{w}'^{k} \\ \mathbf{w}'^{\ell} \end{bmatrix}; \text{ and } \mathbf{N}_{u} \equiv \begin{bmatrix} \mathbf{N}_{u}^{0} & \mathbf{N}_{u}^{0k} & \mathbf{N}_{u}^{0\ell} \\ \mathbf{O} & \mathbf{N}_{u}^{k} & \mathbf{O} \\ \mathbf{O} & \mathbf{O} & \mathbf{N}_{u}^{\ell} \end{bmatrix} \tag{2.61}$$

in which, $n = n^{0} + n^{k} + n^{\ell}$ is the total number of bodies in the tree-type system. Note here that like serial systems, Eq. (2.52), the matrix, $\mathbf{N}_{u}$, for tree-type systems, Eq. (2.61), can also be proven to be the transpose of the matrix $\mathbf{N}_{\ell}$ given by Eq. (2.23).

2.6 Algorithm for Constraint Wrenches

The problem undertaken here as given the motion (joint trajectories) of an open-loop serial or tree-type system, compute the joint reactions, i.e., constraint wrenches, and the driving torques/forces required to achieve the given motion. Based on the formulations given in Sections 2.4 and 2.5, a recursive algorithm to compute the quantities is presented here for the following input for $i=1, \ldots, n$

1. Constant Denavit-Hartenberg (DH) parameters (see Appendix A for definitions), a_{i}, b_{i}, and α_{i}, for a revolute joint; and a_{i}, b_{i}, and θ_{i}, for a prismatic joint

2. Time history of the variable DH parameter and its first and second derivatives, i.e., θ_i, $\dot{\theta}_i$, $\ddot{\theta}_i$ for a revolute joint; and b_i, $\dot{b}_i$, $\ddot{b}_i$ for a prismatic joint.
3. Mass of ith link, m_i
4. Inertia tensor of the ith link about O_i in the the frame attached to it, i.e., $\mathcal{F}_{i+1}$.
5. Vector, $\mathbf{r}_i$, denoting the position of joint O_{i+1} from the mass center, C_i, in $\mathcal{F}_{i+1}$.
6. External wrench, $\mathbf{w}_i^e$, if any on ith link acting at O_i.

The motion of the base is also assumed known, i.e., $\mathbf{t}_0$ and $\dot{\mathbf{t}}_0$ are given. In case the base is fixed, $\mathbf{t}_0 = \dot{\mathbf{t}}_0 = \mathbf{0}$. The algorithm then computes the joint reactions and the driving torques/forces using a two stage recursions, i.e., first the twist and twist rates are computed from the base body to the tip body, followed by the wrench calculations from the tip body to the base body.

Algorithm 2.1: Constraint Wrench for Serial Systems
Recursive stage I: $\mathbf{t}_0 = \mathbf{0}$, $\dot{\mathbf{t}}_0 = \mathbf{0}$ (Assuming base body fixed)

$$For \quad i = 1 \quad \cdots \quad n \tag{2.62}$$
$$\mathbf{t}_i = \mathbf{A}_{i,i-1}\mathbf{t}_{i-1} + \mathbf{p}_i\dot{\theta}_i$$
$$\dot{\mathbf{t}}_i = \mathbf{A}_{i,i-1}\dot{\mathbf{t}}_{i-1} + \dot{\mathbf{A}}_{i,i-1}\mathbf{t}_{i-1} + \mathbf{p}_i\ddot{\theta}_i + \dot{\mathbf{p}}_i\dot{\theta}_i$$
$$end \ loop$$

Recursive stage II:

$$For \quad i = n \quad \cdots \quad 1 \tag{2.63}$$
$$\mathbf{w}_i^* = \mathbf{M}_i\dot{\mathbf{t}}_i + \mathbf{W}_i\mathbf{M}_i\mathbf{t}_i$$
$$\mathbf{w}_{i-1,i} = \mathbf{A}'_{i,i+1}\mathbf{w}_{i,i+1} + \mathbf{w}_i^* - \mathbf{w}_i^e$$
$$\tau_i = \mathbf{p}_i^T\mathbf{w}_{i-1,i}$$
$$end \ loop$$

Algorithm 2.2: Constraint Wrench for Tree-type Systems
Recursive stage I:

For main chain   (2.64a)

$$\mathbf{t}_0 = \mathbf{0}, \quad \dot{\mathbf{t}}_0 = \mathbf{0}$$

For $i = 1 \quad \cdots \quad n^0$

$$\mathbf{t}_i = \mathbf{A}_{i,i-1}\mathbf{t}_{i-1} + \mathbf{p}_i\dot{\theta}_i$$

$$\dot{\mathbf{t}}_i = \mathbf{A}_{i,i-1}\dot{\mathbf{t}}_{i-1} + \dot{\mathbf{A}}_{i,i-1}\mathbf{t}_{i-1} + \mathbf{p}_i\ddot{\theta}_i + \dot{\mathbf{p}}_i\dot{\theta}_i$$

end loop

For any subchain, $j = k, \ell$ (2.64b)

$$\mathbf{t}_0 = \mathbf{t}_j^0 , \quad \dot{\mathbf{t}}_0 = \dot{\mathbf{t}}_j^0$$

For $i = 1 \quad \cdots \quad n^j$

$$\mathbf{t}_i = \mathbf{A}_{i,i-1}\mathbf{t}_{i-1} + \mathbf{p}_i\dot{\theta}_i$$

$$\dot{\mathbf{t}}_i = \mathbf{A}_{i,i-1}\dot{\mathbf{t}}_{i-1} + \dot{\mathbf{A}}_{i,i-1}\mathbf{t}_{i-1} + \mathbf{p}_i\ddot{\theta}_i + \dot{\mathbf{p}}_i\dot{\theta}_i$$

end loop

Recursive Stage II:

For any subchain, $j = k, \ell$ (2.65a)

For $i = n^j \quad \cdots \quad 1$

$$\mathbf{w}_i^* = \mathbf{M}_i\dot{\mathbf{t}}_i + \mathbf{W}_i\mathbf{M}_i\mathbf{t}_i$$

$$\mathbf{w}_{i-1,i} = \mathbf{A}'_{i,i+1}\mathbf{w}_{i,i+1} + \mathbf{w}_i^* - \mathbf{w}_i^e$$

$$\tau_i = \mathbf{p}_i^T\mathbf{w}_{i-1,i}$$

end loop

For main chain $\hspace{8cm}$ (2.65b)

For $\quad i = n^0 \quad \cdots \quad 1$

$\mathbf{w}_i^* = \mathbf{M}_i \dot{\mathbf{t}}_i + \mathbf{W}_i \mathbf{M}_i \mathbf{t}_i$

$\begin{cases} if \qquad i \neq j \\[2mm] \mathbf{w}_{i-1,i} = \mathbf{A}'_{i,i+1} \mathbf{w}_{i,i+1} + \mathbf{w}_i^* - \mathbf{w}_i^e \\[2mm] else \\[2mm] \mathbf{w}_{j-1,j} = \mathbf{A}'_{j,j+1} \mathbf{w}_{j,j+1} + \mathbf{A}'^{j}_{0,1} \mathbf{w}^j_{0,1} + \mathbf{w}_j^* - \mathbf{w}_j^e \\[2mm] end \end{cases}$

$\tau_t = \mathbf{p}_i^T \mathbf{w}_{i-1,1}$

end loop

Equations in (2.62) and (2.63) are the recursive form of $\mathbf{t} = \mathbf{N}_l \mathbf{N}_d \dot{\boldsymbol{\theta}}$ and $\tilde{\mathbf{w}} = \mathbf{N}_l^T \mathbf{w}'$ given by Eq. (2.15) and Eq. (2.49), respectively. Similarly, equations in (2.64) and (2.65) are the recursive forms of $\mathbf{t} = \mathbf{N}_l \mathbf{N}_d \dot{\boldsymbol{\theta}}$ and $\tilde{\mathbf{w}} = \mathbf{N}_l^T \mathbf{w}'$ given by Eqs. (2.22) and Eq. (2.60), respectively. Note that the twists and twist rates of the main chain are evaluated first followed by those of subchains, whereas the constraint wrenches are evaluated in the reverse order.

2.7 Applications

The methodology proposed to compute the constraint wrenches is illustrated here using a planar serial two-link manipulator, a planar tree-type gripper, and two spatial serial six-link manipulators.

2.7.1 Two-link manipulator

A two-link serial planar manipulator is shown in Fig. 2.5. Its first body is attached to the base by a revolute joint at its mass center. Similarly, the second body is attached at its mass center to the first one by another revolute joint. The system has two degree of freedom. Thus, the 2-vector of joint rates, $\dot{\boldsymbol{\theta}} \equiv [\dot{\theta}_1 \quad \dot{\theta}_2]^T$ —θ_1 and θ_2 being the joint angles, as indicated in

Fig. 2.5—can be treated as the independent joint rate vector. Computations are now shown below.

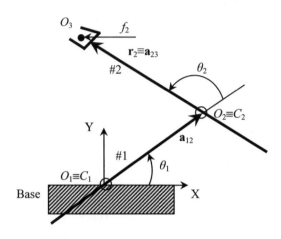

Fig. 2.5. Two-link serial manipulator

. Twist and twist rates
The twists of the bodies, #1 and #2, are determined using Eq. (2.3) as

$$\mathbf{t}_1 = \mathbf{p}_1 \dot{\theta}_1, \text{ and } \mathbf{t}_2 = \mathbf{A}_{21}\mathbf{t}_1 + \mathbf{p}_2 \dot{\theta}_2$$

where the base is fixed, i.e., $\mathbf{t}_0 = \mathbf{0}$. Moreover, the 6-vectors, $\mathbf{p}_1$ and $\mathbf{p}_2$, are associated with the revolute joint axes orthogonal to the XY plane of motion, i.e., $\mathbf{p}_1 = \mathbf{p}_2 \equiv \begin{bmatrix} 0 & 0 & 1 & 0 & 0 & 0 \end{bmatrix}^T$. The 6×6 twist propagation matrix, $\mathbf{A}_{21}$, is then given using Eq. (2.4) as

$$\mathbf{A}_{21} \equiv \begin{bmatrix} \mathbf{1} & \mathbf{O} \\ \tilde{\mathbf{a}}_{21} & \mathbf{1} \end{bmatrix}, \text{ where } \mathbf{a}_{21} = -\mathbf{a}_{12} \equiv -\begin{bmatrix} a_{12x} \\ a_{12y} \\ 0 \end{bmatrix}$$

a_{12x} and a_{12y} being the components of vector $\mathbf{a}_{12}$ along X and Y-axes, respectively.   Hence, $\mathbf{t}_1 \equiv \begin{bmatrix} 0 & 0 & \dot{\theta}_1 & 0 & 0 & 0 \end{bmatrix}^T$,

$\mathbf{t}_2 \equiv \begin{bmatrix} 0 & 0 & \dot{\theta}_{12} & -a_{12y}\dot{\theta}_1 & a_{12x}\dot{\theta}_1 & 0 \end{bmatrix}^T$, and

$\dot{\mathbf{t}}_1 \equiv \begin{bmatrix} 0 & 0 & \ddot{\theta}_1 & 0 & 0 & 0 \end{bmatrix}^T$,

$\dot{\mathbf{t}}_2 \equiv \begin{bmatrix} 0 & 0 & \ddot{\theta}_{12} & -(a_{12y}\ddot{\theta}_1 + \dot{a}_{12y}\dot{\theta}_1) & (a_{12x}\ddot{\theta}_1 + \dot{a}_{12x}\dot{\theta}_1) & 0 \end{bmatrix}^T$,

where $\dot{\theta}_{12} \equiv \dot{\theta}_1 + \dot{\theta}_2$ and $\ddot{\theta}_{12} \equiv \ddot{\theta}_1 + \ddot{\theta}_2$.

- Joint wrenches

Using Eq. (2.49), joint wrenches are calculated as

$$\begin{bmatrix} \mathbf{w}_{01} \\ \mathbf{w}_{12} \end{bmatrix} = \begin{bmatrix} 1 & \mathbf{A}'_{12} \\ \mathbf{O} & 1 \end{bmatrix} \begin{bmatrix} \mathbf{w}'_1 \\ \mathbf{w}'_2 \end{bmatrix}$$

which is evaluated inward from the last link, link #2, i.e,

$$\mathbf{w}_{12} = \mathbf{w}'_2 = \mathbf{w}^*_2 - \mathbf{w}^e_2 \text{ ; where } \mathbf{w}^e_2 = \begin{bmatrix} \tilde{\mathbf{r}}_2 \mathbf{f}^e_2 \\ \mathbf{f}^e_2 \end{bmatrix}$$

and the external force, $\mathbf{f}^e_2 \equiv [-f_2 \quad 0 \quad 0]^T$, which is acting at O_3. The vector, $\mathbf{r}_2$, denoting the point O_3 from C_2 is defined as, $\mathbf{r}_2 \equiv [r_{2x} \quad r_{2y} \quad 0]^T$. Hence, $\mathbf{w}^e_2 \equiv [0 \quad 0 \quad r_{2y} f_2 \quad -f_2 \quad 0 \quad 0]^T$. Now, the inertia wrench of link 2 is obtained as, $\mathbf{w}^*_2 \equiv \mathbf{M}_2 \mathbf{t}_2 + \mathbf{W}_2 \mathbf{M}_2 \mathbf{E}_2 \mathbf{t}_2$, where the 6×6 mass matrix, $\mathbf{M}_2$, is as follows:

$$\mathbf{M}_2 \equiv \begin{bmatrix} \mathbf{I}_2 & m_2 \tilde{\mathbf{d}}_2 \\ -m_2 \tilde{\mathbf{d}}_2 & m_2 \mathbf{1} \end{bmatrix}$$

In which $\mathbf{d}_2 = \mathbf{0}$ as O_2 and C_2 coincide. Hence, $\mathbf{I}_2 = \mathbf{I}^c_2 - m_2 \tilde{\mathbf{d}}^2_2 = \mathbf{I}^c_2$. Furthermore, the 6×6 angular velocity matrix, $\mathbf{W}_2$, is given by:

$$\mathbf{W}_2 \equiv \begin{bmatrix} \tilde{\boldsymbol{\omega}}_2 & \mathbf{O} \\ \mathbf{O} & \tilde{\boldsymbol{\omega}}_2 \end{bmatrix}, \qquad \text{where} \qquad \boldsymbol{\omega}_2 = [0 \quad 0 \quad \dot{\theta}_{12}]^T. \qquad \text{Therefore,}$$

$$\mathbf{w}^*_2 \equiv [0 \quad 0 \quad I^c_2 \ddot{\theta}_{12} \quad -m_2(a_{12y}\ddot{\theta}_1 + \dot{a}_{12y}\dot{\theta}_1) \quad m_2(a_{12x}\ddot{\theta}_1 + \dot{a}_{12x}\dot{\theta}_1) \quad 0]^T.$$

Now, summing up the above expressions, the constraint wrench at joint O_2 is obtained as

$$\mathbf{w}_{12} = \begin{bmatrix} 0 & 0 & I^c_2 \ddot{\theta}_{12} - r_{2y} f_2 \end{bmatrix}$$

$$-m_2(a_{12y}\ddot{\theta}_1 + \dot{a}_{12y}\dot{\theta}_1) + f_2 \quad m_2(a_{12x}\ddot{\theta}_1 + \dot{a}_{12x}\dot{\theta}_1) \quad 0 \Big]^T$$

where the nonzero components are

$$\tau_{12z} \equiv I^c_2 \ddot{\theta}_{12} - r_{2y} f_2$$

$$f_{12x} \equiv -m_2(a_{12y}\ddot{\theta}_1 + \dot{a}_{12y}\dot{\theta}_1) + f_2$$

$$f_{12y} \equiv m_2(a_{12x}\ddot{\theta}_1 + \dot{a}_{12x}\dot{\theta}_1)$$

Note above that the first expression, τ_{12z}, is nothing but the driving torque, τ_2, i.e., $\tau_{12z} \equiv \tau_2 = \mathbf{p}^T_2 \mathbf{w}_{12}$, whereas f_{12x} and f_{12y} are reactions at joint 2. The constraint wrench at the other joint, i.e., O_1, is obtained next as

$$\mathbf{w}_{01} = \mathbf{w}_1' + \mathbf{A}_{12}' \mathbf{w}_2', \text{ where } \mathbf{A}_{12}' \equiv \begin{bmatrix} 1 & \tilde{\mathbf{a}}_{12} \\ \mathbf{O} & 1 \end{bmatrix}$$

and $\mathbf{A}_{12}' \mathbf{w}_2' \equiv [0 \quad 0 \quad \tau_{12z} - a_{12y} f_{12x} + a_{12x} f_{12y} \quad f_{12x} \quad f_{12y} \quad 0]^T$.

Using, $\mathbf{d}_1 = \mathbf{0}$, $\boldsymbol{\omega}_1 = [0 \quad 0 \quad \dot{\theta}_1]^T$, and $\mathbf{w}_1^e = \mathbf{0}$, the vector, $\mathbf{w}_1'$, is found as $\mathbf{w}_1' = [0 \quad 0 \quad I_1^c \ddot{\theta}_1 \quad 0 \quad 0 \quad 0]^T$. Finally, vector $\mathbf{w}_{01}$ is obtained as

$$\mathbf{w}_{01} = [0 \quad 0 \quad I_1^c \ddot{\theta}_1 + \tau_{12z} - a_{12y} f_{12x} + a_{12x} f_{12y} \quad f_{12x} \quad f_{12y} \quad 0]^T$$

whose nonzero components are

$$\tau_{01z} \equiv I_1^c \ddot{\theta}_1 + \tau_{12z} - a_{12y} f_{12x} + a_{12x} f_{12y}$$

$$f_{01x} \equiv f_{12x}$$

$$f_{01y} \equiv f_{12y}$$

where τ_{01z} is nothing but the driving torque at joint O_1, τ_1, i.e., $\tau_{01z} \equiv \tau_1 = \mathbf{p}_1^T \mathbf{w}_{01}$. Now, noting the following from Fig. 2.5

$$a_{12x} = a_{12} \cos \theta_1; \quad a_{12y} = a_{12} \sin \theta_1; \quad r_{2x} = r_2 \cos \theta_{12}; \quad r_{2y} = r_2 \sin \theta_{12},$$

the expressions for the joint torques and forces are finally obtained as

$$\tau_{01z} \equiv \tau_1 = I_1^c \ddot{\theta}_1 + I_2^c \ddot{\theta}_{12} + m_2 r_1^2 \ddot{\theta}_1 - a_{12} f_2 \sin \theta_1 - r_2 f_2 \sin \theta_{12}$$

$$\tau_{12z} \equiv \tau_2 = I_2^c \ddot{\theta}_{12} - r_2 f_2 \sin \theta_{12}$$

$$f_{01x} = f_{12x} = -m_2 (a_{12} \sin \theta_1 \ddot{\theta}_1 + a_{12} \cos \theta_1 \dot{\theta}_1^2) + f_2$$

$$f_{01y} = f_{12y} = m_2 (a_{12} \cos \theta_1 \ddot{\theta}_1 - a_{12} \sin \theta_1 \dot{\theta}_1^2)$$

The above expressions for τ_1 and τ_2 are exactly same as reported in [61]. It is pointed out here that even if a planar problem is solved here, all the vectors and matrices were treated as per the spatial motion instead of their simplified versions indicated in Sections 2.1 and 2.3. This is done to make the understanding of spatial systems clear.

2.7.2 Four link gripper

Figure 2.6 shows a four-link four-DOF planar tree-type gripper whose links are #1^0, #2^0, #3^0 in the main chain, and #1^1 in the subchain, which are coupled with revolute joints, 1^0, 2^0, 3^0, and 1^1. Such system can be used to hold an object, as required in material handling by robots, and others. The constraint wrenches for the gripper is evaluated using the following steps.

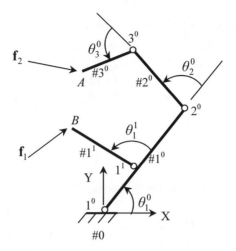

(a) Kinematic diagram

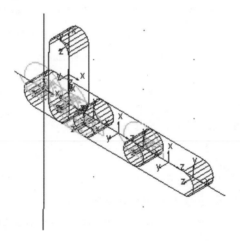

(b) ADAMS model

Fig. 2.6. Four-link tree-type gripper

- Twists

The generalized twist, $\mathbf{t}$, for the system is the 24-vector that is defined as

$$\mathbf{t} \equiv \begin{bmatrix} \mathbf{t}^0 \\ \mathbf{t}^1 \end{bmatrix}, \text{ where } \mathbf{t}^0 \equiv \begin{bmatrix} \mathbf{t}_1^0 \\ \mathbf{t}_2^0 \\ \mathbf{t}_3^0 \end{bmatrix} \text{ and } \mathbf{t}^1 \equiv \mathbf{t}_1^1$$

in which the 18-vector, $\mathbf{t}^0$, is the generalized twist for the main chain, whereas the 6-vector, $\mathbf{t}^1$, is the generalized twist for its subchain. Note that the superscripts, "0" and "1", denote the main chain and its subchain, respectively, whereas the subscripts denote the bodies. Now, the 24×24 and 24×4, DeNOC matrices, $\mathbf{N}_l$ and $\mathbf{N}_d$, respectively, are obtained using Eq. (2.23) as

$$\mathbf{N}_l \equiv \begin{bmatrix} \mathbf{N}_l^0 & \mathbf{O} \\ \mathbf{N}_l^{10} & \mathbf{N}_l^1 \end{bmatrix}; \text{ and } \mathbf{N}_d \equiv \begin{bmatrix} \mathbf{N}_d^0 & \mathbf{O} \\ \mathbf{O} & \mathbf{N}_d^1 \end{bmatrix}$$

The 18×18 matrix, $\mathbf{N}_l^0$, the 6×18 matrix, $\mathbf{N}_l^{10}$, and the 6×6 matrix, $\mathbf{N}_l^1$, are as follows:

$$\mathbf{N}_l^0 \equiv \begin{bmatrix} \mathbf{1} & \mathbf{O} & \mathbf{O} \\ \mathbf{A}_{21}^0 & \mathbf{1} & \mathbf{O} \\ \mathbf{A}_{31}^0 & \mathbf{A}_{32}^0 & \mathbf{1} \end{bmatrix}; \mathbf{N}_l^{01} \equiv \begin{bmatrix} \mathbf{A}_{10}^1 & \mathbf{O} & \mathbf{O} \end{bmatrix}; \text{ and } \mathbf{N}_l^1 \equiv \mathbf{1}$$

where $\mathbf{A}_{21}^0$, $\mathbf{A}_{31}^0$ and $\mathbf{A}_{32}^0$ are the 6×6 twist propagation matrices in the main chain "0" and $\mathbf{A}_{10}^1$ corresponds to subchain 1 which are defined according to Eq. (2.4). Furthermore, the 18×3 matrix, $\mathbf{N}_d^0$, and the 6×1 matrix, $\mathbf{N}_d^1$, are as follow:

$$\mathbf{N}_d^0 \equiv \begin{bmatrix} \mathbf{p}_1^0 & \mathbf{0} & \mathbf{0} \\ \mathbf{0} & \mathbf{p}_2^0 & \mathbf{0} \\ \mathbf{0} & \mathbf{0} & \mathbf{p}_3^0 \end{bmatrix}; \text{ and } \mathbf{N}_d^1 \equiv \mathbf{p}_1^1$$

in which $\mathbf{p}_1^0, \mathbf{p}_2^0$ and $\mathbf{p}_3^0$ are the 6-vectors of joint-rate propagation in the main chain, and $\mathbf{p}_1^1$ corresponds to subchain 1 defined according to Eq. (2.7). Furtheremore, the 4-vector of joint-rates is defined by

$$\dot{\boldsymbol{\theta}} \equiv \begin{bmatrix} \dot{\boldsymbol{\theta}}^0 \\ \dot{\boldsymbol{\theta}}^1 \end{bmatrix}, \text{ where } \dot{\boldsymbol{\theta}}^0 \equiv \begin{bmatrix} \dot{\theta}_1^0 & \dot{\theta}_2^0 & \dot{\theta}_3^0 \end{bmatrix}^T \text{ and } \dot{\boldsymbol{\theta}}^1 \equiv \dot{\theta}_1^1$$

- Joint Wrenches
Using Eq. (2.60)

$$\tilde{\mathbf{w}} = \mathbf{N}_u \mathbf{w}', \text{ where } \tilde{\mathbf{w}} \equiv \begin{bmatrix} \tilde{\mathbf{w}}^0 \\ \tilde{\mathbf{w}}^1 \end{bmatrix}; \mathbf{w}' \equiv \begin{bmatrix} \mathbf{w}'^0 \\ \mathbf{w}'^1 \end{bmatrix}; \text{ and } \mathbf{N}_u \equiv \begin{bmatrix} \mathbf{N}_u^0 & \mathbf{N}_u^{01} \\ \mathbf{O} & \mathbf{N}_u^1 \end{bmatrix}$$

in which, $\widetilde{\mathbf{w}}$ and $\mathbf{w}'$, are the 24-vectors, and $\mathbf{N}_u$ is the 24×24 matrix. Moreover, the vectors and matrices associated with the main chain, 0, are given by

$$\widetilde{\mathbf{w}}^0 \equiv \begin{bmatrix} \mathbf{w}_{01} \\ \mathbf{w}_{12} \\ \mathbf{w}_{23} \end{bmatrix}^0 ; \widetilde{\mathbf{w}}'^0 \equiv \begin{bmatrix} \mathbf{w}'_1 \\ \mathbf{w}'_2 \\ \mathbf{w}'_3 \end{bmatrix}^0 ; \mathbf{N}_u^0 \equiv \begin{bmatrix} \mathbf{1} & \mathbf{A}'_{12} & \mathbf{A}'_{13} \\ \mathbf{O} & \mathbf{1} & \mathbf{A}'_{23} \\ \mathbf{O} & \mathbf{O} & \mathbf{1} \end{bmatrix}^0$$

where 0 on the top-right of the vector and matrix representations denote that all the vectors and matrices belong to the main chain, 0. For the sub-chain 1, Eq. (2.60) yields

$$\widetilde{\mathbf{w}}^1 \equiv \mathbf{w}_{01}^1 \text{ and } \mathbf{N}_u^1 \equiv \mathbf{1}$$

whereas the other vectors and matrices are as follows:

$$\mathbf{N}_u^{01} \equiv \mathbf{N}_u^0 \mathbf{A}_0'^1 \mathbf{N}_{u1}^1, \text{ where } \mathbf{A}_0'^1 \equiv \begin{bmatrix} \mathbf{A}_{01}'^1 \\ \mathbf{O} \\ \mathbf{O} \end{bmatrix}, \text{ and } \mathbf{N}_{u1}^1 \equiv \mathbf{1}$$

Unlike the two link manipulator, further details are avoided here due to complex expressions for the scalar elements of the constraint wrenches. Moreover, a MATLAB program is used to compute the constraint wrenches for the following inputs:

Link length, mass and inertia of each link are taken as 0.05 m, 0.2042 kg, and 2.0800×10^{-4} kg-m^4, respectively. The mass centers of the links are located at mid-point of corresponding links. The link #1^1 is attached at the mid-point of the link #1^0. The joint trajectories for all the joints are taken from [60] as

$$\theta_i = \theta_i(0) + \frac{\theta_i(T) - \theta_i(0)}{T}\left[t - \frac{T}{2\pi}\sin\left(\frac{2\pi}{T}t\right)\right], \text{ for } i=1^0, 2^0, 3^0, \text{ and } 1^1$$

Initial and final configurations are $\theta_i(0) = 0$, $\theta_i(T) = 180^o$, for $i=1^0$, 2^0, 3^0, and $\theta_i(0) = 90^o$, $\theta_i(T) = 270^0$ for $i=1^1$, where T=10 s.

The results are shown in Fig. 2.7, which also closely match with those obtained using a model developed in ADAMS software [123], Fig. 2.6(b).

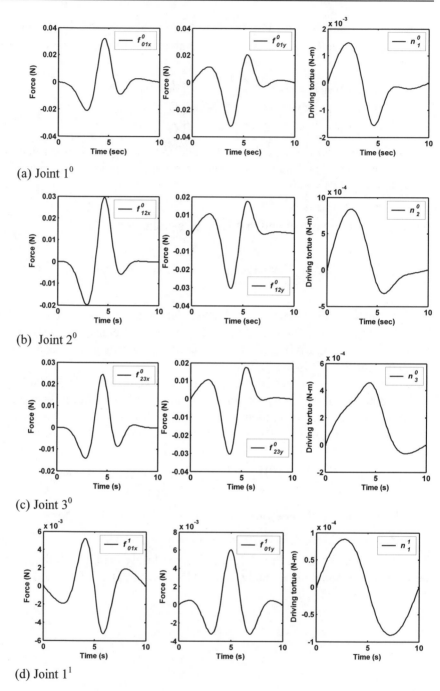

Fig. 2.7. Constraint wrenches for tree-type gripper

2.7.3 Two six-link manipulators

In this subsection, two spatial systems, namely, the six-link manipulators, the Stanford arm and PUMA robot, as shown in Fig. 2.8, are considered here to validate the proposed constraint wrench formulation. The DH parameters, and the mass and inertia properties, which are taken from [79], are given in Table 2.1, where the off-diagonal terms for the inertia tensors, namely, I_{ixy}, I_{iyz}, I_{izx}, are zeros. Note that the elements of the inertia matrix of the ith link are given in its local frame, i.e., the frame attached to the ith link, namely, $\mathcal{F}_{i+1}$ (See Appendix A). In Table 2.1, $r_{i,x}$, $r_{i,y}$ and $r_{i,z}$ denote the components of the vector from C_i to O_{i+1}, i.e., $\mathbf{r}_i$, that are constants in the local frame, $\mathcal{F}_{i+1}$. The joint trajectories are also taken from [79] for the comparison of the results, i.e.,

$$\theta_i = \theta_i(0) + \frac{\theta_i(T) - \theta_i(0)}{T}\left[t - \frac{T}{2\pi}\sin\left(\frac{2\pi}{T}t\right)\right] : \text{For revolute joint}$$

$$b_3 = b_3(0) + \frac{b_3(T) - b_3(0)}{T}\left[t - \frac{T}{2\pi}\sin\left(\frac{2\pi}{T}t\right)\right] : \text{For prismatic joint}$$

For the analysis purposes, the initial and final configurations are taken as
- For Stanford arm

$\theta_i(0) = 0$, for $i \neq 2, 3$, $\theta_2(0) = 90^o$, $b_3(0) = 0$, and $\theta_i(T) = 60^o$, for $i \neq 3$, $b_3(T) = 0.1m$; where T=10 s
- For PUMA robot

$\theta_i(0) = 0$; and $\theta_i(T) = 180^o$; where T=10 s.

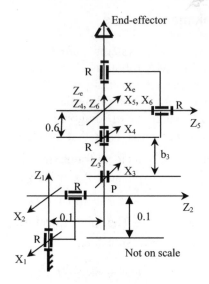

(a) The Stanford arm

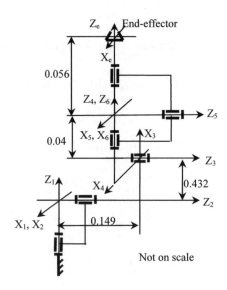

(b) PUMA robot

Fig. 2.8. Six-link manipulators

Table 2.1. DH parameters and inertia properties of six-link manipulators

i	a_i (m)	b_i (m)	α_i (deg)	θ_i (deg)	m_i (kg)	$r_{i,x}$	$r_{i,y}$ (m)	$r_{i,z}$	$I_{i,xx}$	$I_{i,yy}$ (kg-m²)	$I_{i,zz}$
(a) For Stanford arm											
1	0	0.1	-90	θ_1	9	0	-0.1	0	0.01	0.02	0.01
2	0	0.1	-90	θ_2	6	0	0	0	0.05	0.06	0.01
3	0	b_3	0	0	4	0	0	0	0.4	0.4	0.01
4	0	0.6	90	θ_4	1	0	-0.1	0	0.001	0.001	0.0005
5	0	0	-90	θ_5	0.6	0	0	-0.06	0.0005	0.0005	0.0002
6	0	0	0	θ_6	0.5	0	0	-0.2	0.003	0.001	0.002
(b) For PUMA robot											
1	0	0	-90	θ_1	10.521	0	0	0.054	1.612	1.612	0.5091
2	0.432	0.149	0	θ_2	15.761	0.292	0	0	0.4898	8.0783	8.2672
3	0.02	0	90	θ_3	8.767	0.02	0	-0.197	3.3768	3.3768	0.3009
4	0	0.432	-90	θ_4	1.052	0	-0.057	0	0.181	0.1273	0.181
5	0	0	90	θ_5	1.052	0	0	-0.007	0.0735	0.1273	0.0735
6	0	0.056	0	θ_6	0.351	0	0	0.019	0.0071	0.0071	0.0141

The proposed recursive methodology to calculate the constraint wrenches is coded in MATLAB as a general purpose program. The results are shown in Figs. 2.9 and 2.10 for the Stanford arm and PUMA robot, respectively. Note that the Z-component of the constraint moment and force are the driving torque and force for revolute and prismatic joint, respectively, as $[\mathbf{p}_i]_i \equiv [0 \ \ 0 \ \ 1 \ \ 0 \ \ 0 \ \ 0]^T$ for revolute joint, and $[\mathbf{p}_i]_i \equiv [0 \ \ 0 \ \ 0 \ \ 0 \ \ 0 \ \ 1]^T$ for prismatic joint. Hence, the moments and forces of the ith link are evaluated in the ith frame. The driving torque and force, i.e, Z-component of moments in Fig. 2.9 (a), (b), (d-f) and Z-component of the force in Fig. 2.9(c), for the Stanford arm, and the Z-components of the moments in Fis. 2.10(a-f) for the PUMA robot are compared with those in [79]. The results match exactly. Here, however, the interest is also with the reactions, which are not generally reported in the literature. Note also the maximum absolute values of the moments and forces shown in Table 2.2, which in some cases are very high compared to its minimum values.

Table 2.2. Maximum absolute values of the constraint wrenches

Joints	Constraint moment			Constraint force		
	X	Y	Z	X	Y	Z
(a) For Stanford arm						
1	13.23	**24.26**	**0.11**	0.13	0.17	207.08
2	12.90	0.11	22.72	0.15	118.79	0.14
3	0.55	14.68	0.87	59.84	0.12	29.93
4	0.53	14.68	0.87	20.60	0.10	10.30
5	0.62	0.87	1.33	10.79	5.40	8.09
6	0.74	0.98	0.00	4.91	3.68	0.98
(a) For PUMA robot						
1	72.26	69.69	5.41	11.25	9.94	374.97
2	39.44	5.53	78.85	8.17	271.76	11.47
3	2.08	7.66	25.69	116.28	110.09	8.80
4	1.33	9.78	0.12	23.96	2.57	26.05
5	0.17	0.18	0.12	10.50	14.93	10.14
6	0.09	0.13	0.02	3.72	2.54	3.44

Components of moments and forces of the ith link are represented in ith frame

For example, the minimum value of the moment in joint 1 about its Z-axis is only 0.11 Nm, whereas the maximum value about the Y-axis is 24.26 Nm which is 220 times higher than the minimum value. Hence, efforts should be made to design the robot or its trajectory such that the constraint wrenches are equally distributable amongst its joints as much as possible. This way optimum utilization of the resources is possible.

2.8 Summary

In this chapter, using recursive kinematic relations the decoupled natural orthogonal complement (DeNOC) [60] matrices are derived first for serial and tree-type open-loop systems. Next, using the Newton-Euler equations of motion, recursive formulations for the constraint wrenches are proposed. Several planar and spatial systems are analyzed using the proposed methodology, whose results have been verified with those available in the literature or using an alternate model developed in ADAMS software.

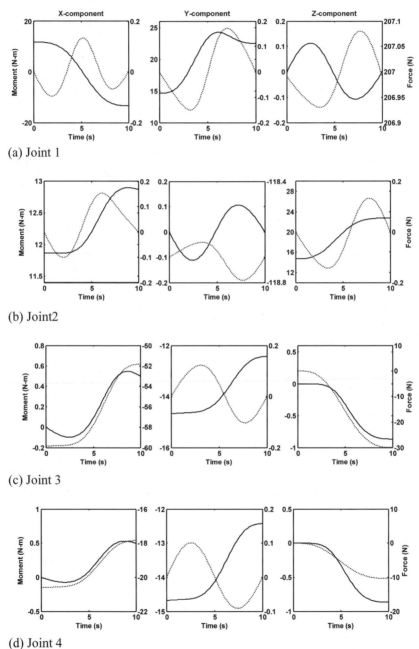

(a) Joint 1

(b) Joint2

(c) Joint 3

(d) Joint 4

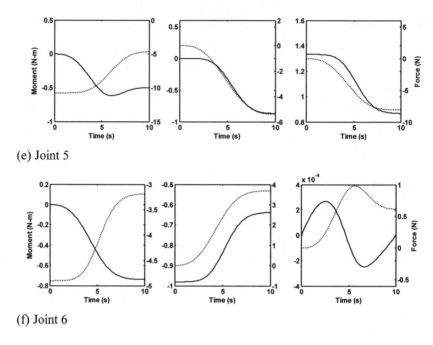

(e) Joint 5

(f) Joint 6

Moment: —— (Left scale of the plots); Force: ⋯⋯⋯ (Right scale of the plots)

Fig. 2.9. Constraint wrenches for the Stanford arm

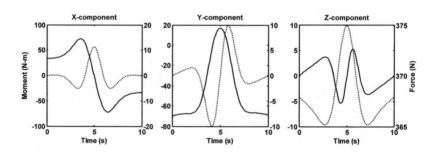

(a) Joint 1

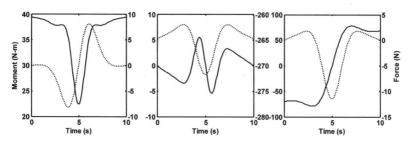

(b) Joint 2

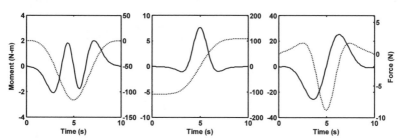

(c) Joint 3

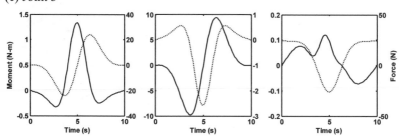

(d) Joint 4

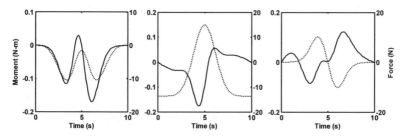

Joint 5

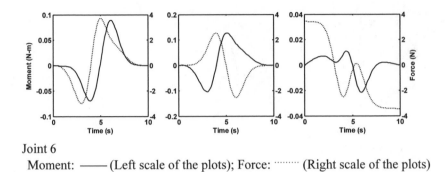

Joint 6
   Moment: ——— (Left scale of the plots); Force: ········· (Right scale of the plots)

Fig. 2.10. Constraint wrenches for PUMA robot

3 Dynamics of Closed-loop Systems

A mechanical system with closed-loop has a circuit of bodies such that the closed circuit can be traced through the system from one body to another and returns to the original body without traversing any body more than once [6]. Kinematic loops occur in many mechanical systems such as vehicles, spacecrafts, and machines for printing, textile and agricultural purposes. In contrast to open-loop systems, as studied in Chapter 2, closed-loop systems tend to possess relatively few degrees of freedom (DOF) compared to the number of connected bodies. Kinematic loops introduce two difficulties: firstly, there is no longer one-to-one correspondence between the joint variables and the motion freedoms of the system due to the loop closure equations; secondly, to define the configuration of the system unambiguously, typically, more number of coordinates than the DOF is used.

In this chapter, dynamic formulation for closed-loop mechanical systems is presented. Traditionally, the closed-loops are cut to make them open. The constraints removed due to the cut joints are retrieved by introducing the Lagrange multipliers. The resulting open system is then called spanning tree. In the next step, identifying the branches of the spanning tree as subsystems, the dynamic formulation of an open system is applied to the subsystems. The computation of all the constraint wrenches at the cut and uncut joints is then done at two levels, namely, at subsystem level, and at body level.

3.1 Equations of Motion

In this section, equations of motion of a closed-loop system are derived based on the philosophy that the closed-loop system is cut open to form an equivalent open-loop system defined as spanning tree.

3.1.1 Spanning tree

Assume that a closed-loop system contains one or more than one closed kinematic loops, as shown in Fig. 3.1. In order to convert such multi-loop system into an equivalent open-loop system, i.e., the spanning tree, the closed kinematic loops are cut at some joints, as indicated in Fig. 3.1. Its equivalent open-loops are shown in Fig. 3.2.

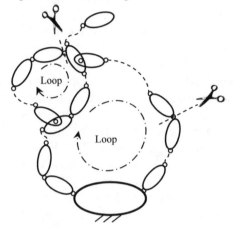

Fig. 3.1. Closed loops in a multibody system

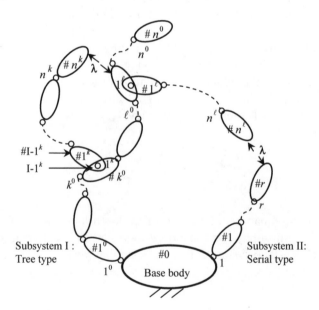

Fig. 3.2. Spanning tree

For a complex multi-loop system, the joints to be cut can be identified using the graph theory [121]. The theory has been successfully used in the past [15-17] to represent a complex mechanical system as a part-contact network. The part-contact network is a directed linear graph representation in which the rigid bodies are represented as nodes, and the joints as edges joining the nodes. The joints to be cut are then decided based on the path whose cumulative degree of freedom (CDOF) is minimum. One such spanning tree thus obtained from the closed-loop system under study is shown in Fig. 3.2.

The branches of the spanning tree which originate from the base body, #0, are referred here as *subsystems*, which could be either serial- or tree-type as defined in Chapter 2. Moreover, the subsystem, e.g. subsystem I of Fig. 3.2, can have branches denoted by superscripts k and ℓ, which are referred here as subchains. For the purpose of defining the architecture of the spanning tree, the fixed body is generally chosen as the base body of the system under study. However, any other floating body whose position, velocity and acceleration are known can also be selected as the base body. The numbering scheme for each subsystem is followed as per the scheme presented in Chapter 2. In addition, the Roman numerals are prefixed before the number of a body to recognize a subsystem to which the body belongs. For example, as indicated in Fig. 3.2, $\#I - 1^k$ denotes the 1st body of the kth subchain in subsystem I, whereas $I - 1^k$ denotes the first joint of the kth subchain in subsystem I. The symbol "#" is used to distinguish the label of bodies from that of joints. The index of subchain is dropped in the serial subsystem, as in subsystem II, because there is no subchain. Assuming that all joints are of one degree-of-freedom, and the total number of bodies is n, then the degree of freedom (DOF) of the spanning tree is given by

$$\text{DOF} = 6 + \sum_{j=I}^{s} n_j \tag{3.1}$$

where n_j is the number of bodies in the jth subsystem with the base body having six DOF. Moreover, s denotes the number of subsystems. For the spanning tree, Fig. 3.2, $n_I = n^0 + n^k + n^\ell$, $n_{II} = r$, and hence, $n = n_I + n_{II}$ and DOF=6+n.

To illustrate the concept of subsystems, consider a planar four-bar mechanism shown in Fig. 3.3. The mechanism has three moving links and one fixed link. Now, one can obtain spanning tree of the four-bar mechanism by cutting its one of the coupler joints. The branches of the spanning tree which originate from the base body, #0, are subsystem I and subsystem II, which are open serial type. Note here that open-loop systems are called by their number of moving links, e.g., two-link manipulator, four-link gripper, etc. as in Chapter 2, whereas the fixed link is also counted in the name of closed-loop systems, e.g., four-bar mechanism. Such terminology is adopted here also as they are commonly known that way in the literature.

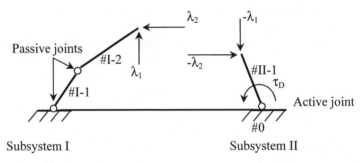

Fig. 3.3. Spanning tree for four-bar mechanism

3.1.2 Determinate and indeterminate subsystems

On a free body of the spanning tree, Fig. 3.2, there may be as many as four categories of moments and forces or wrenches, acting on it, namely, 1) external wrenches from the environment, and those provided by the actuators to drive the system; 2) inertia wrenches due to the motion of the body; 3) the wrenches due to the Lagrange multipliers that represent the reaction moments and forces at the cut joints; and 4) the wrenches due to the reactions moments and forces at the uncut joints.

For a spanning tree resulting from the closed-loop system, if the total number of unknowns, i.e., the driving torques/forces, and the Lagrange multipliers, is equal to the DOF of the system, it is referred as a determinate system. For example, the spanning tree of the four-bar mechanism, as shown in Fig. 3.3, is determinate, as there are three unknowns with a total of three DOF. Based on the above definition, any subsystem originating from the base body of the spanning tree can also be categorized as *determinate*. The determinate subsystem is the one in which the number of unknowns, i.e., the driving torques/forces, and the Lagrange multipliers associated with that subsystem, is equal to its DOF. If the condition for the determinate system is not satisfied then the subsystem is called *indeterminate*. Example of an indeterminate subsystem is subsystem II of the spanning tree resulted from four-bar mechanism shown in Fig. 3.3. It has three unknowns and one DOF. It can, however, be shown that once the unknowns for the determinate subsystem are solved, one or more indeterminate subsystem(s) become determinate. In Fig. 3.3, once λ_1 and λ_2 of determinate subsystem I are solved, subsystem II becomes determinate with one unknown, τ_D, and one-DOF.

3.1.3 The DeNOC matrices for the spanning tree

The serial- and tree-type systems originated from the base body of the spanning tree are the subsystems. Assuming that the base body is fixed, the generalized twist of the jth subsystem is expressed as

$$\mathbf{t}_j = \mathbf{N}_j \dot{\theta}_j, \text{ where } \mathbf{N}_j \equiv (\mathbf{N}_l \mathbf{N}_d)_j \text{ for } j=1, \ldots, s \tag{3.2}$$

where the $6n_j \times n_j$ matrix, $\mathbf{N}_j$, is the natural orthogonal complement (NOC) written as a multiplication of two matrices, namely, the decoupled NOC or DeNOC matrices for the serial or tree-type systems, as derived in Chapter 2. The generalized twist for the spanning tree is then given by

$$\mathbf{t} = \mathbf{N}\dot{\theta} \tag{3.3}$$

where the $6n$-vector, $\mathbf{t}$, the n-vector, $\dot{\theta}$, and the $6n \times n$ matrix, $\mathbf{N}$, are as follows:

$$\mathbf{t} \equiv \begin{bmatrix} \mathbf{t}_1^T & \cdots & \mathbf{t}_s^T \end{bmatrix}^T; \ \dot{\theta} \equiv \begin{bmatrix} \dot{\theta}_1^T & \cdots & \dot{\theta}_s^T \end{bmatrix}^T; \text{ and } \mathbf{N} = diag \begin{bmatrix} \mathbf{N}_1 & \cdots & \mathbf{N}_s \end{bmatrix}$$

3.1.4 Constrained equations of motion for a subsystem

The unconstrained equations of motion, explained in Chapter 2, are used as the starting point to derive the constrained equations of motion for each subsystem of the spanning tree. For a subsystem having n_j moving bodies, the $6n_j$ scalar unconstrained equations of motion are expressed as

$$\mathbf{M}_j \dot{\mathbf{t}}_j + \mathbf{W}_j \mathbf{M}_j \mathbf{E}_j \mathbf{t}_j = \mathbf{w}_j \tag{3.4}$$

where the $6n_j$-vectors $\mathbf{t}_j$, $\dot{\mathbf{t}}_j$, and $\mathbf{w}_j$, respectively, are the generalized twist, twist-rate, and wrench, whereas the $6n_j \times 6n_j$ matrices, $\mathbf{M}_j$, $\mathbf{W}_j$, and $\mathbf{E}_j$, are the generalized mass, angular velocity, and coupling matrices, respectively. For a serial subsystem, the bodies are numbered from 1 to n_j, hence, the $6n_j$-vectors, $\mathbf{t}_j$, $\dot{\mathbf{t}}_j$, and $\mathbf{w}_j$, and the $6n_j{\times}6n_j$ matrices, $\mathbf{M}_j$, $\mathbf{W}_j$, and $\mathbf{E}_j$, are defined as

$$
\mathbf{t}_j \equiv \begin{bmatrix} \mathbf{t}_1 \\ \vdots \\ \mathbf{t}_{n_j} \end{bmatrix} ; \; \dot{\mathbf{t}}_j \equiv \begin{bmatrix} \dot{\mathbf{t}}_1 \\ \vdots \\ \dot{\mathbf{t}}_{n_j} \end{bmatrix} \mathbf{w}_j \equiv \begin{bmatrix} \mathbf{w}_1 \\ \vdots \\ \mathbf{w}_{n_j} \end{bmatrix} \tag{3.5a}
$$

where $\mathbf{M}_j \equiv diag[\mathbf{M}_1 \quad \cdots \quad \mathbf{M}_{n_j}]$; $\mathbf{W}_j \equiv diag[\mathbf{W}_1 \quad \cdots \quad \mathbf{W}_{n_j}]$; and $\mathbf{E}_j \equiv diag[\mathbf{E}_1 \quad \cdots \quad \mathbf{E}_{n_j}]$. For a tree-type subsystem, namely, subsystem I of Fig. 3.2, bodies in the main chain and the subchains are numbered using the strategy explained in Chapter 2. The $6n_j$-vectors, $\mathbf{t}_j$, $\dot{\mathbf{t}}_j$, and $\mathbf{w}_j$, and the $6n_j{\times}6n_j$ matrices, $\mathbf{M}_j$, $\mathbf{W}_j$, and $\mathbf{E}_j$, are defined as

$$
\mathbf{t} \equiv \begin{bmatrix} \mathbf{t}^0 \\ \mathbf{t}^k \\ \mathbf{t}^\ell \end{bmatrix} ; \; \dot{\mathbf{t}} \equiv \begin{bmatrix} \dot{\mathbf{t}}^0 \\ \dot{\mathbf{t}}^k \\ \dot{\mathbf{t}}^\ell \end{bmatrix} ; \; \mathbf{w} \equiv \begin{bmatrix} \mathbf{w}^0 \\ \mathbf{w}^k \\ \mathbf{w}^\ell \end{bmatrix} \tag{3.5b}
$$

where $\mathbf{M}_j \equiv diag[\mathbf{M}^0 \quad \mathbf{M}^k \quad \mathbf{M}^\ell]$; $\mathbf{W}_j \equiv diag[\mathbf{W}^0 \quad \mathbf{W}^k \quad \mathbf{W}^\ell]$; and $\mathbf{E}_j \equiv diag[\mathbf{E}^0 \quad \mathbf{E}^k \quad \mathbf{E}^\ell]$, in which the components of the vectors and matrices are of the sizes according to the number of bodies in the main chain and subchains of the tree-type subsystem. For example, $\mathbf{t}^0$ and $\mathbf{M}^0$ corresponding to the main chain, 0, are the $6n^0$-vector and $6n^0{\times}6n^0$ matrix, respectively.

According to the categorization of the wrenches of the ith body, the wrench, $\mathbf{w}_i$, is composed of the wrench, $\mathbf{w}_i^e$, of the externally applied moments and forces on the ith body by the environment and the driving actuators, the wrench, $\mathbf{w}_i^\lambda$, of the Lagrange multipliers at the cut joints, and the constraint wrench, $\mathbf{w}_i^c$, of the nonworking moments and forces at the uncut joints, i.e., $\mathbf{w}_i \equiv \mathbf{w}_i^e + \mathbf{w}_i^\lambda + \mathbf{w}_i^c$. Hence, Eq. (3.4) can be rewritten as

$$
\mathbf{M}_j \dot{\mathbf{t}}_j + \mathbf{M}_j \mathbf{W}_j \mathbf{t}_j = \mathbf{w}_j^e + \mathbf{w}_j^\lambda + \mathbf{w}_j^c \tag{3.6}
$$

where $\mathbf{w}_j^e$, $\mathbf{w}_j^\lambda$, and $\mathbf{w}_j^c$ denote the $6n_j$-vectors of corresponding moments and forces associated to the jth subsystem. It can be shown that the pre-multiplication of the transpose of the matrix, $\mathbf{N}_j$, with the unconstrained NE equations of motion, Eq. (3.6), yields a set of n_j constrained equations of motion free from the constraint wrenches at the uncut joints, i.e.,

$$\mathbf{N}_j^T (\mathbf{M}_j \dot{\mathbf{t}}_j + \mathbf{W}_j \mathbf{M}_j \mathbf{E}_j \mathbf{t}_j) = \mathbf{N}_j^T (\mathbf{w}_j^e + \mathbf{w}_j^\lambda) \qquad (3.7)$$

Note that the constraint wrenches at the uncut joints do not perform any work. Hence, $\mathbf{N}_j^T \mathbf{w}_j^c = \mathbf{0}$ [68] and vanished from Eq. (3.7). Using the notation for the inertia wrench of the jth subsystem as $\mathbf{w}_j^*$, i.e.,

$\mathbf{M}_j \dot{\mathbf{t}}_j + \mathbf{W}_j \mathbf{M}_j \mathbf{E}_j \mathbf{t}_j \equiv \mathbf{w}_j^*$, Eq. (3.7) is rewritten as

$$\mathbf{N}_j^T \mathbf{w}_j^* = \boldsymbol{\tau}_j^e + \boldsymbol{\tau}_j^\lambda \qquad (3.8)$$

where

$\boldsymbol{\tau}_j^e \equiv \mathbf{N}_j^T \mathbf{w}_j^e$: the n_j-vector of generalized forces due to external wrenches, and those resulting from the actuators, gravity, and dissipation;

$\boldsymbol{\tau}_j^\lambda \equiv \mathbf{N}_j^T \mathbf{w}_j^\lambda$: the n_j-vector of generalized forces due to the Lagrange multipliers.

Equation (3.8) is a set of algebraic equations linear in Lagrange multipliers and the driving torques/forces associated with the jth subsystem. The number of these constrained dynamic equations of motion, Eq. (3.8), i.e., n_j, is certainly less than the $6n_j$ unconstrained NE equations of motion, Eq. (3.4). For a determinate subsystem, n_j is equal to the number of unknown Lagrange multipliers and the driving torques/forces, if any, and can be solved uniquely. With the unknowns solved for the determinate subsystems, some or all of the indeterminate subsystems become determinate, and the process can be continued. This is referred here as subsystem level recursion. Next, to obtain the constraint wrenches at the uncut joints, the recursive algorithm for the open-loop system is used as given in Chapter 2.

To illustrate the above scheme, consider the planar four-bar mechanism shown in Fig. 3.3. Two constrained equations of motion for subsystem I, which is determinate with two unknowns and two DOF, are solved first for the unknown Lagrange multipliers, λ_1 and λ_2. This is followed by the solution of the remaining unknown, τ_D, of subsystem II, which becomes determinate with λ_1 and λ_2 known. Such formulation is expected to increase the efficiency of computation, particularly, when such computations have to be repeated several hundred or thousand times, as in the constraint force optimization problem taken up in Chapters 5 and 6. This is mainly due to the reduction of the dimension of the problem of simultaneous equations.

3.1.5 Constrained equations of motion for the spanning tree

For the spanning tree, Fig. 3.2, the constrained equations are obtained using Eq. (3.8) as

$$\mathbf{N}_j^T \mathbf{w}_j^* = \boldsymbol{\tau}_j^e + \boldsymbol{\tau}_j^\lambda \text{ for } j\text{=I, } \dots , s \qquad (3.9)$$

where s is the number of subsystems in the spanning tree. Note that the subsystems are either a serial- or tree-type originated from the base body, #0, as indicated in Fig. 3.2. Equation (3.9) is written in a compact form as:

$$\mathbf{N}^T \mathbf{w}^* = \boldsymbol{\tau}^e + \boldsymbol{\tau}^\lambda \qquad (3.10)$$

where

$$\mathbf{w}^* \equiv \begin{bmatrix} \mathbf{w}_I^* \\ \vdots \\ \mathbf{w}_s^* \end{bmatrix} ; \boldsymbol{\tau}^e \equiv \begin{bmatrix} \boldsymbol{\tau}_I^e \\ \vdots \\ \boldsymbol{\tau}_s^e \end{bmatrix} ; \text{ and } \boldsymbol{\tau}^\lambda \equiv \begin{bmatrix} \boldsymbol{\tau}_I^\lambda \\ \vdots \\ \boldsymbol{\tau}_s^\lambda \end{bmatrix} \qquad (3.11)$$

$\mathbf{N}$ being the $6n{\times}n$ matrix defined in Eq. (3.3), whereas $\mathbf{w}^*$ is the $6n$-vector, and $\boldsymbol{\tau}^e$, and $\boldsymbol{\tau}^\lambda$ are the n-vectors—$n \equiv n_I + \dots + n_S$ is the total number of bodies in the spanning tree. For a determinate spanning tree resulting from a closed-loop system, the total number of unknowns, i.e., the driving torques/forces and the Lagrange multipliers, is equal to the total number of bodies or the DOF of the system. Hence, the equations of motion given by Eq. (3.10) can be solved simultaneously using any standard method such as LU decomposition and others [122], Such approach to compute the Lagrange multipliers and driving torques/forces is termed here as the system approach, in which the constraint wrenches at the uncut joints are computed at body level recursively. Similar methodology is reported in [67, 72, 132], however, the one presented in Subsection 3.1.4 is developed here.

3.2 Algorithm for Constraint Wrenches

Based on the dynamic analyses presented in Chapter 2 and Section 3.1, the methods to the compute constraint wrenches are categorized as follows:

1. *Traditional method*: In this method, the constraint wrenches, i.e., moments and forces, and driving torques/forces at all the joints are all solved simultaneously using the $6n$ unconstrained NE equations of motion, Eq. (3.6), for j=I, ... , s. For example, using this method nine equations need to be solved simultaneously for the planar four-bar mechanism shown in Fig. 3.3.

2. *System approach*: In this method, all the Lagrange multipliers and the driving torques/forces are solved simultaneously followed by the recursive calculation of the constraint wrenches at the uncut joints. For the four-bar mechanism, Fig. 3.3, three constrained equations for the spanning tree are solved simultaneously, which typically requires order $(3^3/3)$, i.e., order (9) computational complexity. This is followed by the recursive calculation of the constraint wrenches at the three uncut joints using the algorithm presented in Section 2.6.

3. *Subsystem approach*: This method is developed here. It is a two-level recursive method to compute all the Lagrange multipliers, driving torques/forces, and the constraints wrenches at the uncut joints. This is explained in Fig. 3.4. In the first or subsystem level recursion, the determinate subsystems, as defined in Subsection 3.1.2, are identified. The sets of Lagrange multipliers, and the driving torques/forces, if any, of the determinate systems are calculated. With the known Lagrange multipliers of the determinate subsystems, some or all of the remaining indeterminate subsystems become determinate for which the above step is repeated. With all the Lagrange multipliers, and the driving torques/forces of the spanning tree known, the constraint wrenches at the uncut joints of the subsystems are determined using body-level recursion, as presented in Section 2.6. For the four-bar mechanism, Fig. 3.3, two Lagrange multipliers, λ_1 and λ_2, are first solved simultaneously using the constrained equations for subsystem I. This requires order $(2^3/3)$, i.e., order (2.7) computaions. It is followed by the solution of one unknown, τ_D, using only constrained equation for subsystem II, which requires only order (2) computaions. Hence, the total computational complexity to find λ_1, λ_2 and τ_D using the subsystem approach is approximately order (4). This is certainly less than what is requreed using the system approach. Finally, the constraint wrenches at the uncut joints are solved using the body-level recursion.

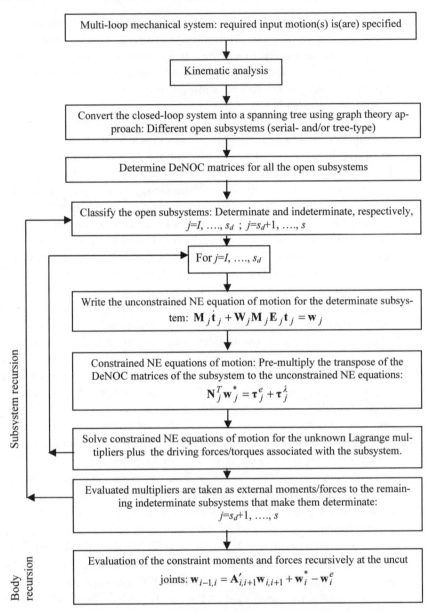

Fig. 3.4. Two-level recursive algorithm to computate of constraint wrenches

The advantages of the proposed two-level recursive subsystem approach are summarized below:

- For a large number of bodies with many kinematic loops, and repeated calculations, the algorithm is proved to be efficient for the computation of constraint wrenches at all joints. For repetitive calculations, the authors used the proposed subsystem approach in the balancing of shaking force and shaking moment where the dynamic algorithm is repeated several times, as done in Chapters 5 and 6. The computational aspects are reported in Tables 3.3 and 3.5.

- It can be used without any modification for the determination of the driving torques/forces, as required in the inverse dynamics of a system necessary for control purposes. Here, the last step of finding the constraint wrenches at the uncut joints need not be carried out.

- Since the constraint wrenches are available at the subsystem level, it may prove useful to analyze them from mechanical design point of view, as the subsystem level results could provide the clue on how it affects the overall system.

3.3 Four-bar Mechanism

Four-bar mechanisms are widely used in mechanical devices owing to their simplicity, ease of manufacturing, and low cost. In this section, a four-bar mechanism is analyzed using the methodology described in Section 3.2. One such four bar mechanism is shown in Fig. 3.5

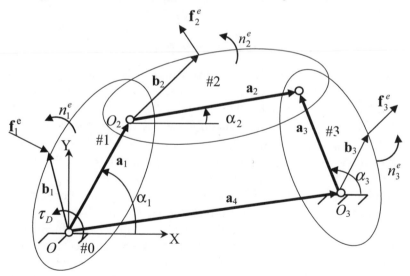

Fig. 3.5. A four-bar mechanism

3.3.1 Equations of motion

The Newton-Euler (NE) equations of motion for the ith rigid link in the fixed inertial frame, OXY, using Eq. (2.37), are written as follows:

$$\mathbf{M}_i \dot{\mathbf{t}}_i + \mathbf{C}_i \mathbf{t}_i = \mathbf{w}_i \qquad (3.12)$$

where the 3×3 matrices, $\mathbf{M}_i$ and $\mathbf{C}_i$ are defined in Eq. (2.39) for planar motion as:

$$\mathbf{M}_i \equiv \begin{bmatrix} I_i & -m_i \mathbf{d}_i^T \overline{\mathbf{E}} \\ m_i \overline{\mathbf{E}} \mathbf{d}_i & m_i \mathbf{1} \end{bmatrix} \text{ and } \mathbf{C}_i \equiv \begin{bmatrix} 0 & \mathbf{0}^T \\ -m_i \omega_i \mathbf{d}_i & \mathbf{O} \end{bmatrix} \qquad (3.13)$$

Equation (3.12) for all the three moving links of the mechanism, i.e., i=1, 2, 3, are put together as

$$\mathbf{M}\dot{\mathbf{t}} + \mathbf{C}\mathbf{t} = \mathbf{w} \qquad (3.14)$$

where the 9×9 matrices, $\mathbf{M}$ and $\mathbf{C}$, are the generalized mass and the matrix of connective inertia terms, respectively, which are defined as

$$\mathbf{M} \equiv diag[\mathbf{M}_1, ..., \mathbf{M}_3] \text{ and } \mathbf{C} \equiv diag[\mathbf{C}_1, ..., \mathbf{C}_3] \qquad (3.15)$$

In Eq. (3.14), the 9-vectors of twist-rate and wrench, $\dot{\mathbf{t}}$ and $\mathbf{w}$, respectively, are

$$\dot{\mathbf{t}} \equiv \begin{bmatrix} \dot{\mathbf{t}}_1^T, & \cdots, & \dot{\mathbf{t}}_3^T \end{bmatrix}^T \text{ and } \mathbf{w} \equiv \begin{bmatrix} \mathbf{w}_1^T, & \cdots, & \mathbf{w}_3^T \end{bmatrix}^T \qquad (3.16)$$

Equation (3.14) shows the nine unconstrained NE equations of motion for the three free-bodies of the mechanism. One can solve these nine equations to calculate the nine unknowns, namely, eight components of reactions at the joints, i.e., $\mathbf{f}_{01}$, $\mathbf{f}_{12}$, $\mathbf{f}_{23}$, and $\mathbf{f}_{04}$, and the driving torque, τ_D. Solution to these unknowns simultaneously is computationally costly, particularly, when it has to be repeated hundred or thousand times, e.g., in an optimization code. Hence, the equations of motion are solved using the algorithm proposed in Section 3.2, i.e., the one shown in Fig. 3.4. For this, the four-bar mechanism is opened by cutting a joint, Fig. 3.6. The loop-constraints are then retrieved by applying the Lagrange multipliers, which are nothing but the reactions at the cut-joint. The Lagrange multipliers and the driving torque are then computed using the two-level recursions, which is referred in Section 3.2 as subsystem approach. The steps are outlined as follows:

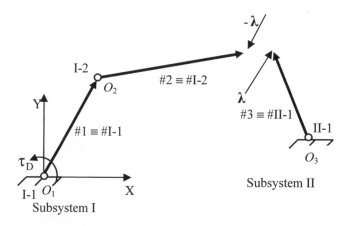

Fig. 3.6. Spanning tree for the four-bar mechanism

1. Classify the spanning tree into subsystems, namely, subsystem I, subsystem II. Note that both the subsystems are indeterminate, i.e., number of associated unknowns is more than the number of equations.

2. Write the unconstrained NE equations of motion for subsystem I:
$$\mathbf{M}\dot{\mathbf{t}} + \mathbf{C}\mathbf{t} = \mathbf{w} \qquad (3.17)$$
where $\mathbf{M} \equiv diag[\mathbf{M}_1, \mathbf{M}_2]$ and $\mathbf{C} \equiv diag[\mathbf{C}_1, \mathbf{C}_2]$ in which $\mathbf{M}_i$ and $\mathbf{C}_i$ for i=1, 2, are given by Eq. (3.13). Moreover, $\mathbf{t} \equiv \begin{bmatrix} \mathbf{t}_1^T & \mathbf{t}_2^T \end{bmatrix}^T$; $\dot{\mathbf{t}} \equiv \begin{bmatrix} \dot{\mathbf{t}}_1^T & \dot{\mathbf{t}}_2^T \end{bmatrix}^T$ and $\mathbf{w} \equiv \begin{bmatrix} \mathbf{w}_1^T & \mathbf{w}_2^T \end{bmatrix}^T$, in which $\mathbf{t}_i \equiv \begin{bmatrix} \omega_i \\ \mathbf{v}_i \end{bmatrix}$ and $\mathbf{w}_i \equiv \begin{bmatrix} n_i \\ \mathbf{f}_i \end{bmatrix}$

3. Convert the unconstrained equations of motion into constrained equations of motion by pre-multiplying the former equations, Eq. (3.17), with the natural orthogonal complement matrix, $\mathbf{N}$, i.e.,
$$\mathbf{N}^T\mathbf{w}^* = \boldsymbol{\tau}^e + \boldsymbol{\tau}^\lambda, \text{ where } \boldsymbol{\tau}^e \equiv \mathbf{N}^T\mathbf{w}^e \text{ and } \boldsymbol{\tau}^\lambda \equiv \mathbf{N}^T\mathbf{w}^\lambda \qquad (3.18)$$
and
$$\mathbf{N} \equiv \begin{bmatrix} \mathbf{p}_1 & \mathbf{0} \\ \mathbf{A}_{21}\mathbf{p}_1 & \mathbf{p}_2 \end{bmatrix}; \quad \mathbf{A}_{21} \equiv \begin{bmatrix} 1 & \mathbf{0}^T \\ \overline{\mathbf{E}}\mathbf{a}_1 & \mathbf{1} \end{bmatrix}; \text{ and } \mathbf{p}_i \equiv \begin{bmatrix} 1 \\ 0 \\ 0 \end{bmatrix} \text{ for } i=1, 2$$

whereas the 2×2 matrix, $\overline{\mathbf{E}}$, is defined in Eq. (2.9b).

4. Determination of $\boldsymbol{\tau}^e$ and $\boldsymbol{\tau}^\lambda$: The wrenches due to external forces and moments at the origin are expressed as follows:

$$\mathbf{w}_1^e \equiv \begin{bmatrix} \mathbf{1} & -\mathbf{b}_1^T \mathbf{E} \\ \mathbf{O} & \mathbf{1} \end{bmatrix} \begin{bmatrix} n_1^e + \tau_D \\ \mathbf{f}_1^e \end{bmatrix} \text{ and } \mathbf{w}_2^e \equiv \begin{bmatrix} \mathbf{1} & -\mathbf{b}_2^T \mathbf{E} \\ \mathbf{O} & \mathbf{1} \end{bmatrix} \begin{bmatrix} n_2^e \\ \mathbf{f}_2^e \end{bmatrix} \quad (3.19)$$

where the 3-vector of generalized torque, $\boldsymbol{\tau}^e \equiv \mathbf{N}^T \mathbf{w}^e$, due to the external forces is obtained as

$$\boldsymbol{\tau}^e = \begin{bmatrix} \tau_D + n_1^e + n_2^e - \mathbf{b}_1^T \overline{\mathbf{E}} \mathbf{f}_1^e - \mathbf{b}_2^T \overline{\mathbf{E}} \mathbf{f}_2^e - \mathbf{a}_1^T \overline{\mathbf{E}} \mathbf{f}_2^e \\ n_2^e - \mathbf{b}_2^T \overline{\mathbf{E}} \mathbf{f}_2^e \end{bmatrix} \quad (3.20)$$

In Fig. 3.6, the Lagrange multiplier, $\boldsymbol{\lambda} \equiv [\lambda_x \quad \lambda_y]^T$, is the 2-vector of the constraint forces at the cut-joint. The wrenches due to the Lagrange multiplier are expressed as follows:

$$\mathbf{w}_1^\lambda = \mathbf{0}; \ \mathbf{w}_2^\lambda \equiv \begin{bmatrix} \mathbf{1} & -\mathbf{a}_2^T \mathbf{E} \\ \mathbf{0} & \mathbf{1} \end{bmatrix} \begin{bmatrix} \mathbf{0} \\ -\boldsymbol{\lambda} \end{bmatrix} \quad (3.21)$$

which leads to expression for the generalized torque due to the Lagrange multipliers, i.e, $\boldsymbol{\tau}^\lambda \equiv \mathbf{N}^T \mathbf{w}^\lambda$, namely,

$$\boldsymbol{\tau}^\lambda = \begin{bmatrix} (\mathbf{a}_1 + \mathbf{a}_2)^T \overline{\mathbf{E}} \boldsymbol{\lambda} \\ \mathbf{a}_2^T \overline{\mathbf{E}} \boldsymbol{\lambda} \end{bmatrix} \quad (3.22)$$

Using the components of the inertia wrench, i.e., n_i^* and $\mathbf{f}_i^*$, and the expressions of $\boldsymbol{\tau}^e$ and $\boldsymbol{\tau}^\lambda$, two scalar constrained equations of motion, Eq. (3.18), are finally obtained as:

$$(\mathbf{a}_1 + \mathbf{a}_2)^T \overline{\mathbf{E}} \boldsymbol{\lambda} - \tau_D = \quad (3.23)$$
$$n_1^* + n_2^* - \mathbf{a}_1^T \overline{\mathbf{E}} \mathbf{f}_2^* - n_1^e - n_2^e + \mathbf{b}_1^T \overline{\mathbf{E}} \mathbf{f}_1^e + \mathbf{b}_2^T \overline{\mathbf{E}} \mathbf{f}_2^e + \mathbf{a}_1^T \overline{\mathbf{E}} \mathbf{f}_2^e$$

$$\mathbf{a}_2^T \overline{\mathbf{E}} \boldsymbol{\lambda} = n_2^* - n_2^e + \mathbf{b}_2^T \overline{\mathbf{E}} \mathbf{f}_2^e \quad (3.24)$$

5. The constrained equations are linear in the components of $\boldsymbol{\lambda}$ and τ_D. However, the system of equations is not solvable uniquely for $\boldsymbol{\lambda}$ and τ_D, as there are two equations and three unknowns. Hence, system I is indeterminate. The steps followed for subsystem I now repeated for the subsystem II to derive its constrained equation.

6. Write the unconstrained NE equation of motion for subsystem, II, i.e.,

$$\mathbf{M}\dot{\mathbf{t}} + \mathbf{C}\mathbf{t} = \mathbf{w} \quad (3.25)$$

where $\mathbf{M} \equiv \mathbf{M}_3$ and $\mathbf{C} \equiv \mathbf{C}_3$ —$\mathbf{M}_3$ and $\mathbf{C}_3$ being given by Eq. (3.13) for $i=3$, Moreover, $\mathbf{t} \equiv \mathbf{t}_3$; $\dot{\mathbf{t}} \equiv \dot{\mathbf{t}}_3$ and $\mathbf{w} \equiv \mathbf{w}_3$ in which

$$\mathbf{t}_3 \equiv \begin{bmatrix} \omega_3 \\ \mathbf{v}_3 \end{bmatrix} \text{ and } \mathbf{w}_3 \equiv \begin{bmatrix} n_3 \\ \mathbf{f}_3 \end{bmatrix}$$

7. Convert Eq. (3.25) into the constrained NE equations of motion, i.e,

$$\mathbf{N}^T \mathbf{w}^* = \boldsymbol{\tau}^e + \boldsymbol{\tau}^\lambda \tag{3.26}$$

where $\mathbf{N} \equiv \mathbf{p}_3$ and $\mathbf{p}_3 = [1 \quad 0 \quad 0]^T$

8. Determination of $\boldsymbol{\tau}^e$ and $\boldsymbol{\tau}^\lambda$: The wrenches due to the external forces and moments, and the Lagrange multipliers are expressed as follows:

$$\mathbf{w}_3^e \equiv \begin{bmatrix} 1 & -\mathbf{b}_3^T \mathbf{E} \\ \mathbf{O} & 1 \end{bmatrix} \begin{bmatrix} n_3^e \\ \mathbf{f}_3^e \end{bmatrix} + \begin{bmatrix} 1 & -\mathbf{a}_3^T \mathbf{E} \\ \mathbf{O} & 1 \end{bmatrix} \begin{bmatrix} 0 \\ \lambda \end{bmatrix} \text{ and} \tag{3.27}$$

$$\mathbf{w}_3^\lambda \equiv \begin{bmatrix} 1 & -\mathbf{a}_3^T \mathbf{E} \\ \mathbf{O} & 1 \end{bmatrix} \begin{bmatrix} 0 \\ \lambda \end{bmatrix}$$

Moreover, the generalized torques, $\tau^e \equiv \mathbf{N}^T \mathbf{w}^e$ and $\tau^\lambda \equiv \mathbf{N}^T \mathbf{w}^\lambda$, are obtained as

$$\tau^e = n_3^e - \mathbf{b}_3^T \overline{\mathbf{E}} \mathbf{f}_3^e ; \quad \tau^\lambda = -\mathbf{a}_3^T \overline{\mathbf{E}} \lambda \tag{3.28}$$

Finally, the only constrained equation is

$$n_3^* = n_3^e - \mathbf{b}_3^T \overline{\mathbf{E}} \mathbf{f}_3^e - \mathbf{a}_3^T \overline{\mathbf{E}} \lambda \tag{3.29}$$

9. The constrained equation, Eq. (3.29), is not solvable for λ, as there is one equation and two unknowns, i.e., the two components of λ.
10. Now the three constrained equations, Eqs. (3.23), (3.24), and (3.29), are solved simultaneously for the three unknowns λ_x, λ_y, and τ_D.
11. Evaluation of the constraint forces recursively at the uncut joints:

$$\mathbf{w}_{i-1,i} = \mathbf{A}'_{i,i+1} \mathbf{w}_{i,i+1} + \mathbf{w}_i^* - \mathbf{w}_i^e \tag{3.30}$$

The recursive equation, Eq. (3.30), for the four-bar mechanism is essentially the force balance equation, as the revolute joints of the mechanism cannot resist any moment. Hence,
For subsystem I:

$$\mathbf{f}_{12} = \mathbf{f}_2^* - \mathbf{f}_2^e + \lambda \text{ and } \mathbf{f}_{01} = \mathbf{f}_{12} + \mathbf{f}_1^* - \mathbf{f}_1^e \tag{3.31}$$

Subsystem II

$$\mathbf{f}_{03} = \mathbf{f}_3^* - \mathbf{f}_3^e - \lambda \tag{3.32}$$

Note that if the joint between links #1 and #2, is opened, Fig. 3.7, instead of the one between links #2 and #3 as in Fig. 3.6, the resulting subsystem I is indeterminate, whereas subsystem II is determinate.

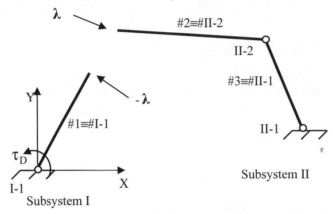

Fig. 3.7. Alternative spanning tree for the four-bar mechanism

Two constrained equations of motion for subsystem II are then solvable for the two unknown components of λ. The solved λ can next be taken as external force to the subsystem I. Hence, one constrained equation of motion for subsystem I is solvable uniquely for the remaining, τ_D. In a similar manner, if the joint between link #0 and #3 is cut the resulting subsystem is only one, which has three unknowns and three DOF. Therefore, the number of subsystem and their nature, i.e., determinate or indeterminate, depends on the joint to be cut. For complex mechanical systems, e.g., vehicles, the concept of graph theory can be used for the determination of which the joints to be cut.

3.3.2 Numerical example

A four-bar mechanism is considered here whose link lengths, masses, and the moment of inertia are given in Table 3.1. They are used to find the constraint forces at the joints of the mechanism. The fixed frame, *XY*, is located at joint 1, Fig. 3.5. The analysis is made by cutting the joint between link #0 and #3. The resulting spanning tree is then a serially connected three-link subsystem having two unknown components of the Lagrange multiplier, namely, λ_x, λ_y, and driving torque, τ_D. The components of the constraint wrenches are now obtained as follows:

1. The compomnets of Lagrange multipliers, λ_x, λ_y, and τ_D, are evaluated first which are plotted in Figs. 3.8 (a) and 3.9.
2. Taking λ_x and λ_y as external forces, the constraint forces of uncut joints are determined, which are shown in Fig. 3.8(b-d).
3. In order to compare the above results a model of the mechanism is developed using the commercial software, MSC.ADAMS 2005 (Automatic Dynamic Analysis of Mechanical Systems) [123]). The plot of τ_D from ADAMS is compared in Fig. 3.9. Even though Only the comparison of the driving torque, τ_D, is shown in Fig. 3.9, all the other forces are also compared which showed close match.

Table 3.1. Link lengths and inertia properties of a four-bar mechanism

Link number	Link length, a_i [m]	Mass of links, m_i [kg]	Moment of inertia about origin O_i, I_i [kg-m^2]	Mass center Location d_i [m] and θ_i [deg]	
#1	1	1	0.5800	0.5	0
#2	2	1.1597	2.1782	1.0	0
#3	3	1.4399	5.5278	1.5	0
#0	3	-	-	-	-

ω_1=1 rad/sec. Link line joining joint 1 and 3 is horizontal

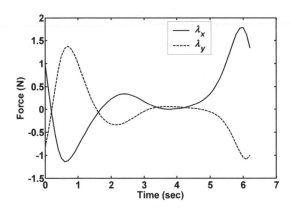

(a) Lagrange multipliers

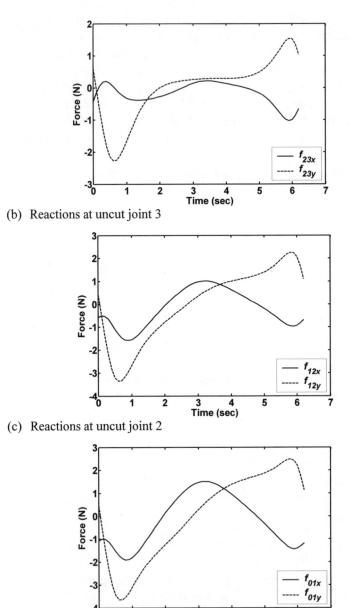

(b) Reactions at uncut joint 3

(c) Reactions at uncut joint 2

(d) Reactions at uncut joint 1

Fig. 3.8. Constraint forces in the four-bar mechanism

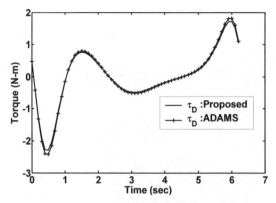

Fig. 3.9. Driving torque and its comparison with ADAMS result

3.4 Carpet Scrapping Machine

The methodology presented in Sections 3.1 and 3.2 to obtain the constraint and driving wrenches is illustrated here with the aid of a carpet scrapping machine developed to clean carpets after it is woven [78]. The machine is shown in Fig. 3.10(a). Two mechanisms, namely, the Hoeken's and the Pantograph, are used in the machine, as shown in Fig. 3.10(b). The Hoeken's mechanism is a crank-rocker mechanism whose coupler generates a partially approximate straight path. The straight line stroke generated by the Hoeken's mechanism is magnified by the Pantograph mechanism. As described in Subsection 3.1.1, the spanning tree of the mechanism given in Fig. 3.10(b) is shown in Fig. 3.11. This is obtained by cutting the appropriate joints. They are between links #1-#2, #2-#5, and #2-#7 of the closed-loops, #0-#1-#2-#3, #0-#1-#2-#5-#4-#0, and #0-#1-#2-#7-#6-#4-#0, respectively. The resulting spanning tree, Fig. 3.11, has three subsystems, I, II, and III. The links and joints of the subsystems are numbered as per the scheme described in subsection 3.1.1. Subsystem I has two moving links, #I-1 and #I-2, with 2-DOF. Subsystem II has only one moving link, #II-1, which is connected to its previous body, i.e., #0, at joint, II-1. Both the subsystems, I and II, are serial types, whereas subsystem III is a tree-type with four moving links numbered as #III-1^0, #III-2^0, #III-3^0, and #III-1^1, which are coupled by four revolute joints denoted as, III-1^0, III-2^0, III-3^0, and III-1^1. Note that each subsystem originates from the base body, #0, which is fixed. Moreover, to avoid clumsiness in Fig. 3.11, the subsystems notations, I, II, and III are not used in the link and joint numbers.

Furthermore, the joint angles θ_1 and θ_2 of subsystem II, θ_1 of subsystem I, and θ_1^0, θ_2^1, θ_3^0, θ_1^1 of subsystem III are treated as generalized coordinates.

(a) Photograph of the machine [78]

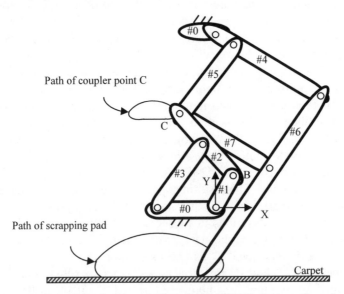

#0-#1-#2-#3: Hoeken's mechanism; #0-#4-#6-#7-#5: Pantograph mechanism

(b) The multiloop scrapping mechanism

Fig. 3.10. Carpet scrapping machine

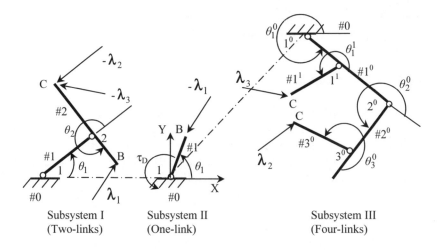

Fig. 3.11. Subsystems of the spanning tree for the carpet scrapping mechanism

The input motion is provided to joint 1 of subsystem II by applying torque τ_D, which needs to be calculated for the known motion of the mechanism. Additionally, three unknown vectors of Lagrange multipliers, λ_i, for $i=1$, 2, 3, as indicated in Fig. 3.11, are

$$\lambda_1 \equiv [\lambda_{1x} \quad \lambda_{1y}]^T, \lambda_2 \equiv [\lambda_{2x} \quad \lambda_{2y}]^T, \text{ and } \lambda_3 \equiv [\lambda_{3x} \quad \lambda_{3y}]^T .$$

Being the motion of the mechanism planar, two components for each Lagrange multiplier represent the reaction forces at the cut revolute joints. Hence, the total number of scalar unknowns are 7, namely, λ_{1x}, λ_{1y}, λ_{2x}, λ_{2y}, λ_{3x}, λ_{3y}, and τ_D. Note that the DOF of the spanning tree (subsystems I, II, and III), is also seven. It implies that the spanning tree is determinate, and seven constrained equations of motion, Eq. (3.10), can be derived at the system level and solved simultaneously for the seven unknowns. The dimension of the associated matrix is 7×7. Alternatively, note that the subsystem, III, has four unknowns, i.e., λ_{2x}, λ_{2y}, λ_{3x}, λ_{3y}, and four-DOF, which allows one to solve for the four unknowns simultaneously at the subsystem level itself using the four constrained equations of motion for subsystem III. The associated matrix size is 4×4. In the next step, evaluated λ_{2x}, λ_{2y}, λ_{3x}, λ_{3y} are taken as known external forces to the subsystem, I, to make it determinate, where λ_{1x}, λ_{1y} are the unknowns and the DOF is two. Hence, the two unknowns can be solved simultaneously at the subsystem level again using the constrained equations of motion for subsystem I. Finally, only τ_D is the unknown in subsystem II, which can be solved by its only constrained equation of motion. Next, the constraint wrenches at the uncut joints are computed recursively at body level. All the steps are explained in detail in the following subsections.

3.4.1 Subsystem III

The subsystem III is four-link subsystem of the spanning tree of the carpet scrapping mechanism. It is a tree-type subsystem. Its longest chain, 1^0-2^0-3^0, is the main chain, whereas the subchain is connected to link #1^0 of the main chain, and has only one link, namely, #1^1. The generalized twist, $\mathbf{t}$, for the subsystem, III, is the 24-vector defined as

$$\mathbf{t} \equiv \begin{bmatrix} \mathbf{t}^0 \\ \mathbf{t}^1 \end{bmatrix}, \text{ where } \mathbf{t}^0 \equiv \begin{bmatrix} \mathbf{t}_1^0 \\ \mathbf{t}_2^0 \\ \mathbf{t}_3^0 \end{bmatrix} \text{ and } \mathbf{t}^1 \equiv \mathbf{t}_1^1 \tag{3.33}$$

in which the 18-vector, $\mathbf{t}^0$, is the generalized twist of the main chain,0, whereas the 6-vector, $\mathbf{t}^1$, is the generalized twist for its subchain, 1. Note that the superscripts, "0" and "1", denote the main chain and its subchain, respectively, whereas the subscripts denote the bodies. Now, the 24×24 and 24×4, DeNOC matrices, $\mathbf{N}_l$ and $\mathbf{N}_d$, respectively, are given by

$$\mathbf{N}_l \equiv \begin{bmatrix} \mathbf{N}_l^0 & \mathbf{O} \\ \mathbf{N}_l^{10} & \mathbf{N}_l^1 \end{bmatrix}; \text{ and } \mathbf{N}_d \equiv \begin{bmatrix} \mathbf{N}_d^0 & \mathbf{O} \\ \mathbf{O} & \mathbf{N}_d^1 \end{bmatrix} \tag{3.34}$$

where the 18×18 matrix, $\mathbf{N}_l^0$, the 6×18 matrix, $\mathbf{N}_l^{10}$, and the 6×6 matrix, $\mathbf{N}_l^1$, are as follows:

$$\mathbf{N}_l^0 \equiv \begin{bmatrix} \mathbf{1} & \mathbf{O} & \mathbf{O} \\ \mathbf{A}_{21}^0 & \mathbf{1} & \mathbf{O} \\ \mathbf{A}_{31}^0 & \mathbf{A}_{32}^0 & \mathbf{1} \end{bmatrix}; \mathbf{N}_l^{10} \equiv \begin{bmatrix} \mathbf{A}_{10}^1 & \mathbf{O} & \mathbf{O} \end{bmatrix}; \text{ and } \mathbf{N}_l^1 \equiv \mathbf{1} \tag{3.35}$$

in which $\mathbf{A}_{21}^0$ $\mathbf{A}_{31}^0$ and $\mathbf{A}_{32}^0$ are the 6×6 twist propagation matrices in the main chain 0, and $\mathbf{A}_{10}^1$ corresponds to subchain 1. Furthermore, the 18×3 matrix, $\mathbf{N}_d^0$, and the 6×1 matrix, $\mathbf{N}_d^1$, are as follow:

$$N_d^0 \equiv \begin{bmatrix} p_1^0 & 0 & 0 \\ 0 & p_2^0 & 0 \\ 0 & 0 & p_3^0 \end{bmatrix}; \text{ and } N_d^1 \equiv p_1^1 \qquad (3.36)$$

In Eq. (3.36), $\mathbf{p}_1^0, \mathbf{p}_2^0$ and $\mathbf{p}_3^0$ are the 6-vectors of joint-rate propagation in the main chain, and $\mathbf{p}_1^1$ corresponds to subchain 1. Finally, the 4-vector of joint-rates for the subsystem, III, is defined by

$$\dot{\boldsymbol{\theta}} = \begin{bmatrix} \dot{\boldsymbol{\theta}}^0 \\ \dot{\boldsymbol{\theta}}^1 \end{bmatrix}, \text{ where } \dot{\boldsymbol{\theta}}^0 \equiv \begin{bmatrix} \dot{\theta}_1^0 & \dot{\theta}_2^0 & \dot{\theta}_3^0 \end{bmatrix}^T \text{ and } \dot{\boldsymbol{\theta}}^1 \equiv \dot{\theta}_1^1 \qquad (3.37)$$

Using Eq. (3.8), the four constrained equations of motion for subsystem III are obtained as

$$\mathbf{N}^T \mathbf{w}^* = \boldsymbol{\tau}^e + \boldsymbol{\tau}^\lambda; \text{ where } \mathbf{N} = \mathbf{N}_I \mathbf{N}_d \qquad (3.38)$$

and $\boldsymbol{\tau}^e \equiv \mathbf{N}^T \mathbf{w}^e$, is the 4-vector of generalized forces due to the external moments and forces. Moreover, $\boldsymbol{\tau}^\lambda \equiv \mathbf{N}^T \mathbf{w}^\lambda$, is the 4-vector of generalized forces due to the Lagrange multipliers associated with the subsystem, III. The 24-vector of generalized wrench, $\mathbf{w}^*$, for the subsystem, III, are now defined as

$$\mathbf{w}^* \equiv \begin{bmatrix} (\mathbf{w}^*)^0 \\ (\mathbf{w}^*)^1 \end{bmatrix}; (\mathbf{w}^*)^0 \equiv \begin{bmatrix} \mathbf{w}_1^* \\ \mathbf{w}_2^* \\ \mathbf{w}_3^* \end{bmatrix}^0; \text{ and } (\mathbf{w}^*)^1 \equiv [\mathbf{w}_1^*]^1 \qquad (3.39)$$

In Eq. (3.39), to avoid clumsiness, "0" at right of the vector representation of $(\mathbf{w}^*)^0$ denotes the superscript "0" for each of its vector entity, i.e., $(\mathbf{w}_i^0)^0$, for i=1, 2, 3. Similarly, one can express the 24-vectors of generalized wrenches, $\mathbf{w}^e$ and $\mathbf{w}^\lambda$, whereas the generalized inertia wrench, $\mathbf{w}^*$, is known from the input motion. Note that λ_2 and λ_3 are acting on links, #3^0 and #1^1, respectively. Hence, the wrenches due to the Lagrange multipliers, $(\mathbf{w}_1^\lambda)^0$, $(\mathbf{w}_2^\lambda)^0$, $(\mathbf{w}_3^\lambda)^0$, and $(\mathbf{w}_1^\lambda)^1$ are as follows:

$$(\mathbf{w}_1^\lambda)^0 = \mathbf{0}, \ (\mathbf{w}_2^\lambda)^0 = \mathbf{0}, \ (\mathbf{w}_3^\lambda)^0 = (\mathbf{A}_{3,C}')^0 \mathbf{w}_{3^c}^\lambda, \text{ and} \qquad (3.40)$$
$$(\mathbf{w}_1^\lambda)^1 = (\mathbf{A}_{1,C}')^1 \mathbf{w}_{1^c}^\lambda$$

where the 6-vectors, $\mathbf{w}_{3c}^{\lambda}$ and $\mathbf{w}_{1c}^{\lambda}$, are the wrenches at cut joints on links, #3^0 and 1^1, i.e., $\mathbf{w}_{3c}^{\lambda} \equiv [0 \quad 0 \quad 0 \quad \lambda_{2x} \quad \lambda_{2y} \quad 0]^T$ and $\mathbf{w}_{1c}^{\lambda} \equiv [0 \quad 0 \quad 0 \quad \lambda_{3x} \quad \lambda_{3y} \quad 0]^T$. Using Eqs. (3.34) and (3.40)

$$\tau^{\lambda} = \mathbf{N}^T \mathbf{w}^{\lambda} = \mathbf{G} \mathbf{w}_c^{\lambda} \tag{3.41}$$

where the 4×12 matrix, $\mathbf{G}$, and the 12-vector, $\mathbf{w}_c^{\lambda}$, are given by

$$\mathbf{G} = \begin{bmatrix} (\mathbf{p}_1^T \mathbf{A}_{1,C}')^0 & (\mathbf{p}_1^T)^0 (\mathbf{A}_{0,C}')^1 \\ (\mathbf{p}_2^T \mathbf{A}_{2,C}')^0 & \mathbf{0}^T \\ (\mathbf{p}_3^T \mathbf{A}_{3,C}')^0 & \mathbf{0}^T \\ \mathbf{0}^T & (\mathbf{p}_1^T \mathbf{A}_{1,C}')^1 \end{bmatrix} \quad \text{and} \quad \mathbf{w}_c^{\lambda} = \begin{bmatrix} \mathbf{w}_{3c}^{\lambda} \\ \mathbf{w}_{1c}^{\lambda} \end{bmatrix} \tag{3.42}$$

Upon simplification of Eqs. (3.41) and (3.42), one can also show that

$$\tau^{\lambda} = \mathbf{J}^T \boldsymbol{\lambda} \tag{3.43}$$

where the 4×4 matrix, $\mathbf{J}$, and the 4-vector, $\boldsymbol{\lambda}$, are as follows:

$$\mathbf{J} \equiv \begin{bmatrix} -a_{1,Cy}^0 & -a_{2,Cy}^0 & -a_{3,Cy}^0 & 0 \\ a_{1,Cx}^0 & a_{2,Cx}^0 & a_{3,Cx}^0 & 0 \\ -a_{0,Cy}^1 & 0 & 0 & -a_{1,Cy}^1 \\ a_{0,Cx}^1 & 0 & 0 & a_{1,Cx}^1 \end{bmatrix}; \quad \text{and} \quad \boldsymbol{\lambda} \equiv \begin{bmatrix} \lambda_{2x} \\ \lambda_{2y} \\ \lambda_{3x} \\ \lambda_{3y} \end{bmatrix} \tag{3.44}$$

in which $a_{i,jx}$ and $a_{i,jy}$ are the components of vector $\mathbf{a}_{i,j}$ along X and Y axes, respectively. The 4×4 matrix, $\mathbf{J}$, is nothing but the Jacobian matrix associated with the kinematic constraints of the subsystem, III, which can be verified as follows: Referring to Fig. 3.11

$$a_{1,2}^0 \cos(\theta_1^0) + a_{2,3}^0 \cos(\theta_1^0 + \theta_2^0) + a_{3,3c}^0 \cos(\theta_1^0 + \theta_2^0 + \theta_3^0) + x_1^0 - x_C^0 = 0 \tag{3.45}$$

$$a_{1,2}^0 \sin(\theta_1^0) + a_{2,3}^0 \sin(\theta_1^0 + \theta_2^0) + a_{3,3c}^0 \sin(\theta_1^0 + \theta_2^0 + \theta_3^0) + y_1^0 - y_C^0 = 0 \tag{3.46}$$

$$a_{0,1}^1 \cos(\theta_1^0) + a_{1,C}^1 \cos(\theta_1^0 + \theta_1^1) + x_1^0 - x_C^1 = 0 \tag{3.47}$$

$$a_{0,1}^1 \sin(\theta_1^0) + a_{1,C}^1 \sin(\theta_1^0 + \theta_1^1) + y_1^0 - y_C^1 = 0 \tag{3.48}$$

where (x_1^0, y_1^0) and (x_C^0, y_C^0) or (x_C^1, y_C^1) are the Cartesian coordinates of joints, 1^0 and C, respectively. Note that, at point C, there are two joints, one along chain "0" and the other along "1". The time derivatives of the constraint equations, namely, Eqs. (3.45-48), give $\mathbf{J}\dot{\boldsymbol{\theta}} = 0$, where $\dot{\boldsymbol{\theta}} \equiv [\dot{\theta}_1^0 \quad \dot{\theta}_2^0 \quad \dot{\theta}_3^0 \quad \dot{\theta}_1^1]^T$, and the expression of the 4×4 matrix, $\mathbf{J}$, is nothing but the one given by Eq. (3.44).

Next, note that the wrenches due to external moments and forces are $(\mathbf{w}_1^e)^0 = (\mathbf{w}_2^e)^0 = (\mathbf{w}_3^e)^0 = (\mathbf{w}_1^e)^1 = \mathbf{0}$, as there is no external moments and forces acting on subsystem III. Hence, the generalized force due to external wrenches is also zero, i.e., $\boldsymbol{\tau}^e = \mathbf{0}$. Finally, the set of four constrained equations of motion, Eq. (3.38), is obtained using the planar motion notations as

$$
\begin{bmatrix}
(n_1^*)^0 + (n_2^*)^0 + (n_3^*)^0 + (n_1^*)^1 - (\mathbf{a}_{12}^T)^0 \overline{\mathbf{E}}(\mathbf{f}_2^*)^0 - (\mathbf{a}_{13}^T)^0 \overline{\mathbf{E}}(\mathbf{f}_3^*)^0 - (\mathbf{a}_{01}^T)^1 \overline{\mathbf{E}}(\mathbf{f}_1^*)^1 \\
(n_2^*)^0 + (n_3^*)^0 - (\mathbf{a}_{23}^T)^0 \overline{\mathbf{E}}(\mathbf{f}_3^*)^0 \\
(n_3^*)^0 \\
(n_1^*)^1
\end{bmatrix} \quad (3.49)
$$
$$
=
\begin{bmatrix}
-a_{1,Cy}^0 & a_{1,Cx}^0 & -a_{0,Cy}^1 & a_{0,Cx}^1 \\
-a_{2,Cy}^0 & a_{2,Cx}^0 & 0 & 0 \\
-a_{3,Cy}^0 & a_{3,Cx}^0 & 0 & 0 \\
0 & 0 & -a_{1,Cy}^1 & a_{1,Cx}^1
\end{bmatrix}
\begin{bmatrix}
\lambda_{2x} \\
\lambda_{2y} \\
\lambda_{3x} \\
\lambda_{3y}
\end{bmatrix}
$$

in which all the vectors, $\mathbf{a}$ and $\mathbf{f}^*$, are the 2-vectors for the planar problem. Moreover, n^* is scalar denoting the inertia moment about the axis orthogonal to the plane of motion. Equation (3.49) has four unknowns, λ_{2x}, λ_{2y}, λ_{3x}, and λ_{3y}, that can be solved simultaneously, say, using LU decomposition of their coefficient matrix $\mathbf{J}^T$ [122]. The constraint forces and moments at the uncut joints, i.e., joints 1^0, 2^0, 3^0, and 1^1, are then easily computed recursively at the body level.

3.4.2 Subsystem I

Considering, λ_{2x}, λ_{2y}, λ_{3x}, and λ_{3y}, computed from the subsystem, III, as known, subsystem I becomes determinate. The unknowns, λ_{1x}, λ_{1y}, associated with the subsystem, I, are now calculated as follows: Subsystem I is a serial-type having two links and originates from joint I-1 of the base. Therefore, the 12×12 and 12×2, DeNOC matrices, $\mathbf{N}_l$ and $\mathbf{N}_d$, respectively, are given by

$$\mathbf{N}_l \equiv \begin{bmatrix} \mathbf{1} & \mathbf{O} \\ \mathbf{A}_{21} & \mathbf{1} \end{bmatrix}; \ \mathbf{N}_d \equiv \begin{bmatrix} \mathbf{p}_1 & \mathbf{0} \\ \mathbf{0} & \mathbf{p}_2 \end{bmatrix} \tag{3.50}$$

in which the 6×6 matrix, $\mathbf{A}_{21}$, and the 6-vectors, $\mathbf{p}_1$ and $\mathbf{p}_2$, are as follows:

$$\mathbf{A}_{21} \equiv \begin{bmatrix} \mathbf{1} & \mathbf{O} \\ \tilde{\mathbf{a}}_{21} & \mathbf{1} \end{bmatrix}; \mathbf{p}_1 \equiv \begin{bmatrix} \mathbf{e}_1 \\ \mathbf{0} \end{bmatrix}; \text{ and } \mathbf{p}_2 \equiv \begin{bmatrix} \mathbf{e}_2 \\ \mathbf{0} \end{bmatrix}$$

where $\tilde{\mathbf{a}}_{21}$ is the skew-symmetric matrix associated with the vector, $\mathbf{a}_{21}$, whereas the 3-vectors, $\mathbf{e}_1$ and $\mathbf{e}_2$, are the unit vectors along the axes of joints 1 and 2, respectively. Using Eq. (3.8), two constrained equations of motion for subsystem I are obtained as

$$\mathbf{N}^T \mathbf{w}^* = \boldsymbol{\tau}^e + \boldsymbol{\tau}^\lambda; \text{ where } \mathbf{N} = \mathbf{N}_l \mathbf{N}_d \tag{3.51}$$

and the terms, $\boldsymbol{\tau}^e \equiv \mathbf{N}^T \mathbf{w}^e$ and $\boldsymbol{\tau}^\lambda \equiv \mathbf{N}^T \mathbf{w}^\lambda$, are the 2-vectors of generalized forces due to external moments and forces, and the Lagrange multipliers associated with subsystem I, respectively. The 12-vector of generalized wrenches, $\mathbf{w}^*, \mathbf{w}^e$, and $\mathbf{w}^\lambda$, for the subsystem, I, are now defined as follows:

$$\mathbf{w}^* \equiv \begin{bmatrix} \mathbf{w}_1^* \\ \mathbf{w}_2^* \end{bmatrix}; \ \mathbf{w}^e \equiv \begin{bmatrix} \mathbf{w}_1^e \\ \mathbf{w}_2^e \end{bmatrix}; \text{ and } \mathbf{w}^\lambda \equiv \begin{bmatrix} \mathbf{w}_1^\lambda \\ \mathbf{w}_2^\lambda \end{bmatrix} \tag{3.52}$$

where

$$\mathbf{w}_1^e = \mathbf{0}, \ \mathbf{w}_2^e = -\mathbf{A}_{2,C}'(\mathbf{w}_{3^c}^\lambda + \mathbf{w}_{1^c}^\lambda), \mathbf{w}_1^\lambda = \mathbf{0}, \text{ and } \mathbf{w}_2^\lambda = \mathbf{A}_{2,B}' \mathbf{w}_B^\lambda \tag{3.53}$$

The 6-vector, $\mathbf{w}_B^\lambda$, being defined as, $\mathbf{w}_B^\lambda \equiv [0 \quad 0 \quad 0 \quad \lambda_{1x} \quad \lambda_{1y} \quad 0]^T$. Using Eq. (3.50)

$$\boldsymbol{\tau}^\lambda = \mathbf{N}^T \mathbf{w}^\lambda = \mathbf{G} \mathbf{w}_B^\lambda \tag{3.54}$$

where the 2×6 matrix, $\mathbf{G}$, is as follows:

$$\mathbf{G} \equiv \begin{bmatrix} \mathbf{p}_1^T \mathbf{A}_{1,B}' \\ \mathbf{p}_2^T \mathbf{A}_{2,B}' \end{bmatrix} \tag{3.55}$$

Upon simplification, Eqs. (3.54) and (3.55) yields

$$\tau^\lambda = \mathbf{J}^T \lambda \tag{3.56}$$

where the 2×2 matrix, $\mathbf{J}$, and 2-vector, λ, are as follows:

$$\mathbf{J} \equiv \begin{bmatrix} -a_{1,By} & -a_{2,By} \\ a_{1,Bx} & a_{2,Bx} \end{bmatrix}; \text{ and } \lambda \equiv \begin{bmatrix} \lambda_{1x} \\ \lambda_{1y} \end{bmatrix} \tag{3.57}$$

The 2×2 matrix, $\mathbf{J}$, is nothing but the Jacobian matrix associated with the kinematic constraints of subsystem I, which can be verified from Fig. 3.11 as

$$a_{12} \cos\theta_1 + a_{2,B} \cos(\theta_1 + \theta_2) - x_B = 0 \tag{3.58}$$

$$a_{12} \sin\theta_1 + a_{2,B} \sin(\theta_1 + \theta_2) - y_B = 0 \tag{3.59}$$

where (x_B, y_B) are the Cartesian coordinates of the point, B. The time derivatives of the constraint equations, Eqs. (3.58-59), give $\mathbf{J}\dot{\theta} = 0$, where $\dot{\theta} \equiv \begin{bmatrix} \dot{\theta}_1 & \dot{\theta}_2 \end{bmatrix}^T$ and the 2×2 matrix, $\mathbf{J}$, is noting but the one shown in Eq. (3.57). Finally, the set of two constrained equations, Eq. (3.51), is given using the planar motion notation as

$$\begin{bmatrix} n_1^* + n_2^* - \mathbf{a}_{12}^T \overline{\mathbf{E}} \mathbf{f}_2^* \\ n_2^* \end{bmatrix} = \begin{bmatrix} -a_{1,By} & a_{1,Bx} \\ -a_{2,By} & a_{2,Bx} \end{bmatrix} \begin{bmatrix} \lambda_{1x} \\ \lambda_{1y} \end{bmatrix} + \begin{bmatrix} \mathbf{a}_{1,C}^T \overline{\mathbf{E}}(\lambda_2 + \lambda_3) \\ \mathbf{a}_{2,C}^T \overline{\mathbf{E}}(\lambda_2 + \lambda_3) \end{bmatrix} \tag{3.60}$$

Equation (3.60) can be solved simultaneously, similar to Eq. (3.49).

3.4.3 Subsystem II

Subsystem II of Fig. 3.11, has only one link coupled with the base body #0 at joint II-1. Hence, its DeNOC matrices, $\mathbf{N}_l$ and $\mathbf{N}_d$, are the 6×6 identity matrix and the 6-vector of joint-rate propagation, respectively, i.e.,

$$\mathbf{N}_l = \mathbf{1} \text{ and } \mathbf{N}_d = \mathbf{p}_1 \tag{3.61}$$

Using Eq. (3.8), one constrained equation of motion for the subsystem II is obtained as

$$\mathbf{N}^T \mathbf{w}^* = \tau^e + \tau^\lambda; \text{ where } \mathbf{N} = \mathbf{N}_l \mathbf{N}_d = \mathbf{p}_1 \tag{3.62}$$

and the scalars, $\tau^e \equiv \mathbf{N}^T \mathbf{w}^e$ and $\tau^\lambda \equiv \mathbf{N}^T \mathbf{w}^\lambda$, are obtained as follows: The 6-vectors of generalized wrenches, $\mathbf{w}^*$, $\mathbf{w}^e$, and $\mathbf{w}^\lambda$, for the subsystem, II, are given by

$$\mathbf{w}^* \equiv \mathbf{w}_1^*; \ \mathbf{w}^e \equiv \mathbf{w}_1^e; \text{ and } \mathbf{w}^\lambda \equiv \mathbf{w}_1^\lambda \tag{3.63}$$

where

$$\mathbf{w}_1^e = -\mathbf{A}'_{1,B}\mathbf{w}_B^\lambda + \mathbf{w}_D^\tau, \text{ and } \mathbf{w}_1^\lambda = \mathbf{0} \tag{3.64}$$

in which $\mathbf{w}_D^\tau \equiv [0 \ \ 0 \ \ \tau_D \ \ 0 \ \ 0 \ \ 0]^T$. Using the above expressions, Eq. (3.62) is rewritten as

$$n_1^* = \mathbf{a}_{1,B}^T \mathbf{E}\boldsymbol{\lambda}_1 + \tau_D \tag{3.65}$$

which has only τ_D unknown that can be easily computed.

3.4.4 Numerical example

The multiloop scrapping mechanism shown in Fig. 3.10(b) has the parameters shown in Table 3.2. They are used here to find the constraint forces at the joints of the mechanism. The input motion provided to link #II-1 is a constant speed of 45 *rpm* (712 *rad/s*). The fixed frame, *XYZ*, is located at joint II-1, Fig. 3.11, where axis Z is orthogonal to the page. Joints I-1 and III-1^0 are located at (-0.089*m*, 0) and (0.038*m*, 0.410*m*), respectively. Joint between #I-1 and #II-2 is located at the mid of link #II-2. Joint III-1^1 is at 0.096*m* on link #III-1^0 from joint III-1^0.

Table 3.2. Link parameters of the scrapping machine

Subsystem	Link	Length [m]	Mass [kg]	Moment of inertia at link origin [kg-m²] ×10^{-3}
I	1	$a_{1,2}$ =0.115	3.0	13.27
	2	$a_{2,B}$=0.115	5.0	22.18
II	1	$a_{1,B}$=0.038	1.5	0.73
III	1^0	a_{12}^0=0.335	2	156.74
	2^0	a_{23}^0=0.239	10.5	2449.10
	3^0	$a_{3,c}^0$=0.239	3.0	57.12
	1^1	$a_{1,c}^1$=0.239	3.0	57.12

Kinematic analysis is performed using the philosophy reported in [61]. On-ly the position analysis results are shown in Fig. 3.12. The detailed deriva-tions of the kinematics are avoided here because the focus of this book is on the constraint wrench computation. The components of the constraint wrenches are now obtained as follows:

1. The Lagrange multipliers of subsystem III, namely, λ_{2x}, λ_{2y}, λ_{3x}, and λ_{3y}, are evaluated first using Eq. (3.49), which are plotted in Fig 3.13.

2. Taking λ_{2x}, λ_{2y}, λ_{3x}, and λ_{3y} as external forces, the Lagrange multiplies in the subsystem, I, namely, λ_{1x} and λ_{1y}, are evaluated in the next step using Eq. (3.60), as shown in Fig. 3.14.

3. Finally, τ_D is evaluated for subsystem II using Eq. (3.65). In order to compare this result the carpet scrapping mechanism is also modelled using the commercial software, MSC.ADAMS 2005 (Automatic Dynamic Analysis of Mechanical Systems) [123] , as shown in Fig. 3.15(a). The plots of τ_D from Eq. (3.65) and that of ADAMS are shown in Fig. 3.15(b). Even though Only the comparison of the driving torque, τ_D, is shown in Fig. 3.15(b), all other forces are also compared which showed close match.

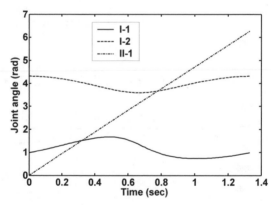

(a) Subsystems I and II

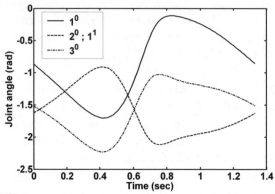

(b) Subsystem III

Fig. 3.12. Joint angles for carpet scrapping mechanism

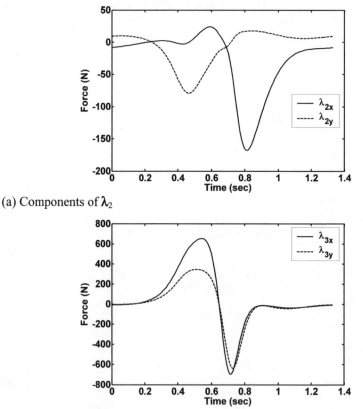

(a) Components of λ_2

(b) Components of λ_3

Fig. 3.13. Lagrange multipliers in subsystems III

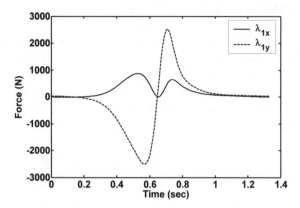

Fig. 3.14. Lagrange multipliers in subsystem I

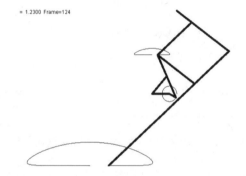

(a) ADAMS model of the carpet scrapping machine

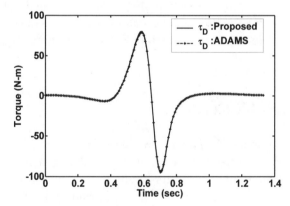

(b) Comparison of driving torque

Fig. 3.15. ADAMS model and variation of driving torque for the carpet scrapping mechanism

3.4.5 Computation efficiency

In this subsection, a look into the computational efficiency of the proposed two-level recursive algorithm is given. Table 3.3 shows the comparison of the theoretical order of computations and the time taken by the CPU of a Pentium IV computer. Three methods, as described in Section 3.2, are considered, namely, the traditional method using the 21×21 matrix, the system approach using the 7×7 matrix and the recursive calculations, and the two-level recursive subsystem approach using the 4×4, 2×2, 1×1 matrices and the recursive calculations. The CPU time is estimated for 134 positions of the mechanism during one complete rotation of the link, #II-1.

Table 3.3. CPU time for carpet scrapping mechanism in Pentium IV(2.60 GHz)

Methods	Theoretical order of computations	CPU time in sec
Traditional (matrix size: 21×21)	$O(21^3/3)=O(3087)$	0.219
System approach (matrix size: 7×7)	$O(7^3/3)+O(7\times2)=O(128.3)$	0.156 $(30.59)^+$
Subsystem approach (matrix sizes: 4×4, 3×3, 1×1)	$O(4^3/3+3^3/3+1)+O(7\times2)=O(45.3)$	0.156 (30.59)

* For simulation time of 1.34 sec (i.e., time period of the crank) with step size of 0.01 sec. + Percentage savings over the traditional method

A significant reduction in the CPU time is observed in both the system and subsystem approaches over the traditional method. However, there is no significant difference in the savings between system and subsystem approaches. This is mainly due to the planar nature of the system at hand, because the spatial four-bar mechanism in Section 3.5 shows significant difference.

3.5 Spatial RSSR Mechanism

The constraint wrench formulation presented in Sections 3.1 and 3.2 is now applied to a three-dimensional spatial problem, namely, the RSSR mechanism, where R and S stand for Revolute and Spherical joints, respectively. The mechanism is shown in Fig. 3.16, whose kinematic equivalent [80] is given Fig. 3.17, where the spherical joint between link #3 and #6 is considered equivalent to three revolute joints intersecting at a point, whereas the second spherical joint between links #1 and #3 is substituted with two intersecting revolute joints instead of three. This is due to the fact that the rotation of the coupler link, #3, about its own axis does not affect the overall input-output motion of the mechanism. Hence, the rotation is redundant and the associated DOF is removed from the equivalent mechanism. As a result, the equivalent mechanism is a 7R mechanism whose joint 7 is assumed to be driven by an actuator. The links are numbered as #0, …, #6, —#0 being the base link, which is fixed. The DH notations defined in Appendix A are now used to define its architecture as

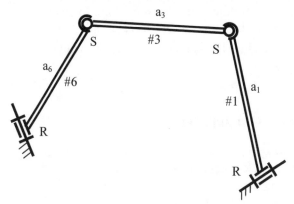

Fig. 3.16. The RSSR mechanism

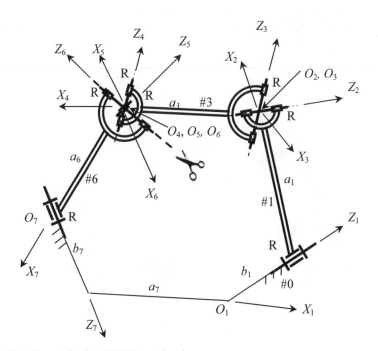

Fig. 3.17. The equivalent RSSR mechanism

$$a_2 = a_4 = a_5 = 0; \quad b_2 = b_3 = b_4 = b_5 = b_6 = 0; \tag{3.66}$$
$$\alpha_2 = \alpha_4 = \alpha_5 = 0$$

The 3-vector, $\mathbf{a}_{i,i+1}$, is then given by

$$\mathbf{a}_{i,i+1} = b_i \mathbf{e}_i + a_i \mathbf{x}_{i+1} \tag{3.67}$$

where $\mathbf{e}_i$ and $\mathbf{x}_{i+1}$ are the unit 3-vectors along the axes, Z_i and X_{i+1}, respectively. Note that the unit vectors have simple form in their body fixed frame, i.e., ith and $(i+1)$st frame, respectively. They are $[\mathbf{e}_i]_i = [0,0,1]^T$ and $[\mathbf{x}_{i+1}]_{i+1} = [1,0,0]^T$.

3.5.1 Subsystem approach

In this section, the proposed constraint wrench formulation is illustrated with the spatial 7R mechanism. The closed-loop mechanism is first made open by cutting the joint between links #3 and #6 of the RSSR mechanism. In the 7R mechanism, it is assumed that the joint, 6, is to be cut such that one of the subsystems becomes determinate. The cutting line is indicated in Fig. 3.17 by a scissor symbol. The resulting spanning tree has two serial-type subsystems, as shown in Fig. 3.18. They are: subsystem I with one moving link #6, and subsystem II with five serially connected moving links, #1,…., #5.

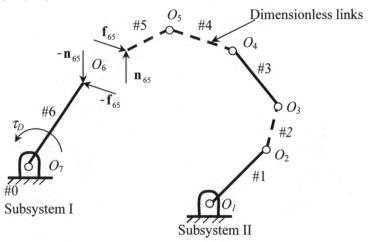

Fig. 3.18. Spanning tree of the 7R mechanism

Since the cut-joint, 6, is a revolute joint there are five components for the vector of Lagrange multipliers, $\boldsymbol{\lambda}$, two corresponding to the reaction moments, λ_1, λ_2, and three to the reaction forces, λ_3, λ_4, λ_5, i.e., $\boldsymbol{\lambda} \equiv [\lambda_1 \quad \lambda_2 \quad \lambda_3 \quad \lambda_4 \quad \lambda_5]^T$. Referring to Fig. 3.18, the reaction moment, $\mathbf{n}_{65}$, and the reaction force, $\mathbf{f}_{65}$, at the cut joint in Frame 6 are given by

$$[\mathbf{n}_{65}]_6 = [\lambda_1 \quad \lambda_2 \quad 0]^T \text{ and } [\mathbf{f}_{65}]_6 = [\lambda_3 \quad \lambda_4 \quad \lambda_5]^T \tag{3.68}$$

It is pointed out here that the subsystem, II, has 5 unknowns, namely, λ_1, ...,λ_5, which is equal to its DOF. It implies that subsystem II is determinate and its five constrained equations of motion in the form of Eq. (3.8) can be solved simultaneously for the five unknown multipliers. The constrained equations of motion, Eq. (3.8), for subsystem II are expressed as

$$\mathbf{N}^T \mathbf{w}^* = \boldsymbol{\tau}^e + \boldsymbol{\tau}^\lambda \tag{3.69}$$

where, $\boldsymbol{\tau}^e \equiv \mathbf{N}^T \mathbf{w}^e$, is the 5-vector of the generalized forces due to the external wrenches and $\boldsymbol{\tau}^\lambda \equiv \mathbf{N}^T \mathbf{w}^\lambda$ is the 5-vector of the generalized forces due to the Lagrange multipliers. The matrix $\mathbf{N}$ is derived next for subsystem II with five moving links. The 30-vector of generalized twist, $\mathbf{t} \equiv [\mathbf{t}_1^T \quad \cdots \quad \mathbf{t}_5^T]^T$, is given as

$$\mathbf{t} = \mathbf{N}\dot{\boldsymbol{\theta}}, \text{ where } \mathbf{N} \equiv \mathbf{N}_l \mathbf{N}_d \tag{3.70}$$

The 30×30 lower block triangular matrix, $\mathbf{N}_l$, and the 30×5 block diagonal matrix, $\mathbf{N}_d$, are the DeNOC matrices, which are given by:

$$\mathbf{N}_l \equiv \begin{bmatrix} \mathbf{1} & \mathbf{O} & \cdots & \mathbf{O} \\ \mathbf{A}_{21} & \mathbf{1} & \cdots & \mathbf{O} \\ \vdots & \vdots & \ddots & \vdots \\ \mathbf{A}_{51} & \mathbf{A}_{52} & \cdots & \mathbf{1} \end{bmatrix}; \mathbf{N}_d \equiv \begin{bmatrix} \mathbf{p}_1 & \mathbf{0} & \cdots & \mathbf{0} \\ \mathbf{0} & \mathbf{p}_2 & \cdots & \mathbf{0} \\ \vdots & \vdots & \ddots & \vdots \\ \mathbf{0} & \mathbf{0} & \cdots & \mathbf{p}_5 \end{bmatrix} \tag{3.71}$$

In Eq. (3.71), the 6×6 twist propagation matrix, $\mathbf{A}_{ij}$, and the 6-vector of joint-rate propagation, $\mathbf{p}_i$, are defined as

$$\mathbf{A}_{ij} \equiv \begin{bmatrix} \mathbf{1} & \mathbf{O} \\ \tilde{\mathbf{a}}_{ij} & \mathbf{1} \end{bmatrix}; \mathbf{p}_i \equiv \begin{bmatrix} \mathbf{e}_i \\ \mathbf{0} \end{bmatrix} \tag{3.72}$$

in which $\mathbf{1}$ and $\mathbf{O}$ are the 3×3 identity and zero matrices, and $\mathbf{0}$ is the 3-vector of zeros. They should be understood as of compatible sizes based on where they appear. The 30-vector of the generalized wrench associated with the Lagrange multipliers, $\mathbf{w}^\lambda$, for subsystem II is represented as

$$\mathbf{w}^\lambda \equiv \left[(\mathbf{w}_1^\lambda)^T, ..., (\mathbf{w}_5^\lambda)^T \right]^T \qquad (3.73)$$

where $\mathbf{w}_i^\lambda = \mathbf{0}$, for $i=1, ..., 4$, as the joints 1, ..., 4 are not cut. Furthermore, the wrench on link 5, $\mathbf{w}_5^\lambda$, due to the Lagrange multipliers at the cut joint, 6, with respect to its origin point, O_5, is given by

$$\mathbf{w}_5^\lambda \equiv \mathbf{A}_{56}' \mathbf{w}_{6^c}^\lambda, \text{ where } \mathbf{A}_{56}' \equiv \begin{bmatrix} \mathbf{1} & \tilde{\mathbf{a}}_{56} \\ \mathbf{O} & \mathbf{1} \end{bmatrix} \text{ and } \mathbf{w}_{6^c}^\lambda \equiv \begin{bmatrix} \mathbf{n}_{65} \\ \mathbf{f}_{65} \end{bmatrix} \qquad (3.74)$$

Moreover, the 6×6 matrix, $\mathbf{A}_{56}'$, is the wrench propagation matrix. The generalized force due to the Lagrange multipliers, $\boldsymbol{\tau}^\lambda$, is then expressed as

$$\boldsymbol{\tau}^\lambda = \mathbf{N}^T \mathbf{w}^\lambda = \mathbf{G} \mathbf{w}_{6^c}^\lambda, \text{ where } \mathbf{G} \equiv \begin{bmatrix} \mathbf{p}_1^T \mathbf{A}_{16}' \\ \mathbf{p}_2^T \mathbf{A}_{26}' \\ \mathbf{p}_3^T \mathbf{A}_{36}' \\ \mathbf{p}_4^T \mathbf{A}_{46}' \\ \mathbf{p}_5^T \mathbf{A}_{56}' \end{bmatrix} \qquad (3.75)$$

In deriving Eq. (3.75), the following properties associated with the twist and wrench propagation matrices are used:

$$\mathbf{A}_{i,j}^T = \mathbf{A}_{j,i}', \text{ and } \mathbf{A}_{i,j}' \mathbf{A}_{j,k}' = \mathbf{A}_{i,k}' \qquad (3.76)$$

Now, the inertia wrenches for the links 2, 4, 5 are zero, as they are massless with no dimensions, i.e.,

$$\mathbf{w}_i^* = \mathbf{0}, \text{ for } i=2, 4, \text{ and } 5 \qquad (3.77)$$

Hence, the expression on the left hand side of Eq. (3.69) is given by

$$\mathbf{N}^T \mathbf{w}^* \equiv \begin{bmatrix} \mathbf{p}_1^T \mathbf{w}_1^* + \mathbf{p}_1^T \mathbf{A}_{13}' \mathbf{w}_3^* \\ \mathbf{p}_2^T \mathbf{A}_{23}' \mathbf{w}_3^* \\ \mathbf{p}_3^T \mathbf{w}_3^* \\ 0 \\ 0 \end{bmatrix} \qquad (3.78)$$

Also, there is no external wrench on subsystem II, hence, the generalized force, $\boldsymbol{\tau}^e \equiv \mathbf{0}$. Upon substitution of Eqs. (3.75) and (3.78) into Eq. (3.69) yields

$$
\begin{bmatrix} \mathbf{p}_1^T \mathbf{w}_1^* + \mathbf{p}_1^T \mathbf{A}_{13}' \mathbf{w}_3^* \\ \mathbf{p}_2^T \mathbf{A}_{23}' \mathbf{w}_3^* \\ \mathbf{p}_3^T \mathbf{w}_3^* \\ 0 \\ 0 \end{bmatrix} = \begin{bmatrix} \mathbf{p}_1^T \mathbf{A}_{16}' \\ \mathbf{p}_2^T \mathbf{A}_{26}' \\ \mathbf{p}_3^T \mathbf{A}_{36}' \\ \mathbf{p}_4^T \mathbf{A}_{46}' \\ \mathbf{p}_5^T \mathbf{A}_{56}' \end{bmatrix} \mathbf{w}_{6^c}^\lambda
\tag{3.79}
$$

where the last two scalar equations are as follows:

$$
\mathbf{p}_4^T \mathbf{A}_{46}' \mathbf{w}_{6^c}^\lambda = 0 \quad \text{and} \quad \mathbf{p}_5^T \mathbf{A}_{56}' \mathbf{w}_{6^c}^\lambda = 0
\tag{3.80-81}
$$

Equations (3.80) and (3.81) are simplified using the system architecture, Eq. (3.66), as

$$
\mathbf{e}_4^T \mathbf{n}_{65} = 0 \quad \text{and} \quad \mathbf{e}_5^T \mathbf{n}_{65} = 0
\tag{3.82-83}
$$

In addition, joint 6 is a revolute joint, which cannot resist any moment about its own axis. Hence,

$$
\mathbf{e}_6^T \mathbf{n}_{65} = 0
\tag{3.84}
$$

In Eqs. (3.82-84), the 3-vectors, $\mathbf{e}_4$, $\mathbf{e}_5$, and $\mathbf{e}_6$, are the unit vectors along axes Z_4, Z_5, and Z_6, respectively, that are orthogonal to each other. Hence, to satisfy the three scalar equations given by Eqs. (5.82-84), the following must be true

$$
\mathbf{n}_{65} = \mathbf{0}
\tag{3.85}
$$

This result is as expected becuase the combination of the three joints, 4, 5, and 6, form a spherical joint, which cannot transmit any moment. The first three scalar equations of Eq. (3.79) are now simplified as

$$
\mathbf{e}_1^T (\mathbf{n}_1^* + \mathbf{n}_3^* + \tilde{\mathbf{a}}_{13} \mathbf{f}_3^*) = \mathbf{e}_1^T \tilde{\mathbf{a}}_{16} \mathbf{f}_{65}
\tag{3.86}
$$

$$
\mathbf{e}_2^T (\mathbf{n}_3^* + \tilde{\mathbf{a}}_{23} \mathbf{f}_3^*) = \mathbf{e}_2^T \tilde{\mathbf{a}}_{26} \mathbf{f}_{65}
\tag{3.87}
$$

$$
\mathbf{e}_3^T \mathbf{n}_3^* = \mathbf{e}_3^T \tilde{\mathbf{a}}_{36} \mathbf{f}_{65}
\tag{3.88}
$$

which leads to the solution for the three reaction components of $\mathbf{f}_{65}$. The reaction forces at the other joints, 1, ..., 5, can then be calculated recursively using the methodology given in Chapter 2.

Knowing the vector of Lagrange multipliers, λ, from subsystem II, six unknowns of subsystem I, i.e., the three components each for the moment, $\mathbf{n}_{76}$, and force, $\mathbf{f}_{76}$, can be solved easily. Note here that the Z-component of $\mathbf{n}_{76}$ in frame $O_7X_7Y_7Z_7$ is nothing but the driving torque at joint 7. It is pointed out here again that the actual RSSR mechanism has two degree of freedom (DOF), including the rotation of the coupler about its own axis. This is the redundant DOF, which introduces centrifugal forces unless the mass center of the coupler is located on its central axis. This makes both the kinematics and dynamics of the equivalent 7R mechanism and the original RSSR mechanism same. The original RSSR mechanism has a total of 18 unknowns, namely, 3 constraint forces for each spherical joint and 5 constraint forces and moments for each revolute joint, and two driving torques. Hence, the problem of the actual RSSR mechanism can be solved using 18 equations of motion for the three links, i.e., #1, #3, and #6. Alternatively, the 7R mechanism has 36 unknowns, i.e., 5 for each revolute joint and one for driving torque. Hence, performing the dynamic analysis of the 7R mechanism in a straightforward manner will be extremely uneconomical. The equations of motion are reduced from 36 to 18 using the special arrangement and dimensions of the links, #2, #4 and #5, of the equivalent 7R mechanism. These links have zero mass and zero length, i.e., $\mathbf{w}_2^* = \mathbf{0}$, $\mathbf{w}_4^* = \mathbf{0}$, and $\mathbf{w}_5^* = \mathbf{0}$, which means, $\mathbf{w}_{12}^* = \mathbf{w}_{23}^*$ and $\mathbf{w}_{34} = \mathbf{w}_{45} = \mathbf{w}_{56}$. Moreover, the joints 3, 4, and 5 form a spherical joint. Hence, $\mathbf{n}_{65} = \mathbf{0}$, as proved in this section. Similarly, the joints, 2 and 3, form a Hook's joint that can resist only moment in the direction of $\mathbf{e}_2 \times \mathbf{e}_3$ which is denoted with $\mathbf{n}_{23}$ and expressed as $\mathbf{n}_{23} = \dfrac{\mathbf{e}_2 \times \mathbf{e}_3}{|\mathbf{e}_2 \times \mathbf{e}_3|} n_{23}$ — n_{23} being the magnitude of $\mathbf{n}_{23}$, i.e., $\mathbf{n}_{23} = |n_{23}|$. As result, the effective number of unknowns are 18, i.e, two components each for $[\mathbf{n}_{01}]_1 = [n_{01x} \quad n_{01y} \quad 0]^T$ and $[\mathbf{n}_{06}]_7 = [n_{06x} \quad n_{06y} \quad 0]^T$, the magnitude of $\mathbf{n}_{23}$, i.e., n_{23}, three components each for, $\mathbf{f}_{01}$, $\mathbf{f}_{23}$, $\mathbf{f}_{56}$, and $\mathbf{f}_{06}$, and one driving torque, τ_D.

3.5.2 Numerical example

The DH parameters, and the mass and inertia properties of the 7R mechanism are shown in Table 3.4. The mass center location and the elements of the inertia tensor for each link are given in their local coordinate frames. The input motion provided to link #6 as a constant speed of 100 rpm (10.472 rad/sec). All the vectors are evaluated in the fixed frame, $X_1 Y_1 Z_1$. Kinematic analysis is performed using the philosophy reported in [80]. Only the joint angles, i.e., θ_1, θ_2, and θ_3, which are required here to perform the dynamic analysis, are plotted in Fig. 3.19. Other joint angles, θ_4, and θ_5 do not play any role in dynamics, as the links #4 and #5 are massless. Hence, the angles, θ_4 and θ_5, are not plotted. The constraint wrenches are now obtained as follows:

1. The Lagrange multipliers of subsystem II, namely, λ_3, λ_4, and λ_5, are evaluated first using Eqs. (3.86-88), which are shown in Fig. 3.20. Note that, $\mathbf{n}_{65} = \mathbf{0}$, hence, λ_1 and λ_2 are also zero.
2. Taking λ_3, λ_4, and λ_5, as external forces in subsystem I, the driving torque is computed which is plotted in Fig. 3.21.

The results are also compared with those obtained from the commercial software, MSC.ADAMS 2005 [123] shown in Fig. 3.22. The results match exactly for λ_3, λ_4, and λ_5. The results for τ_D do not match exactly, mainly, due to the modeling error in the calculation of the generalized inertia tensor of the system at hand.

Table 3.4. DH parameters and mass and inertia properties for the 7R mechanism

Link i	a_i (m)	b_i (m)	α_i (rad)	θ_i (rad)	m_i (kg)	$r_{i,x}$ (m)	$r_{i,y}$	r_{iz}	$I_{i,xx}$ (kg-m^2)×10^{-6}	$I_{i,yy}$	$I_{i,zz}$
0	1.30	0.40	$5\pi/6$	θ_0	-	-	-	-	-	-	-
1	1.00	0.55	$\pi/2$	θ_1	0.084	-0.50	0	0	1.3874	82.080	82.067
2	0	0	$\pi/2$	θ_2	0	0	0	0	0	0	0
3	1.10	0	$\pi/2$	θ_3	0.092	-0.55	0	0	1.5175	106.954	106.954
4	0	0	$\pi/2$	θ_4	0	0	0	0	0	0	0
5	0	0	$\pi/2$	θ_5	0	0	0	0	0	0	0
6	0.50	0	$\pi/2$	θ_6	0.045	-0.25	0	0	0.7374	12.939	12.939

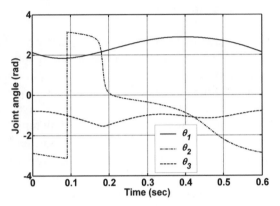

Fig. 3.19. Joint angles for 7R mechanism

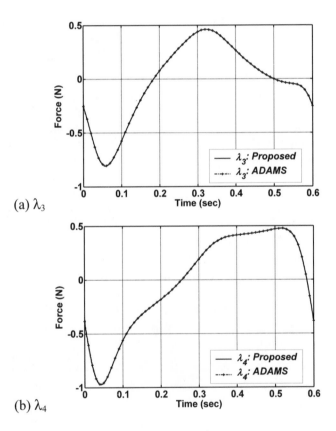

(a) λ_3

(b) λ_4

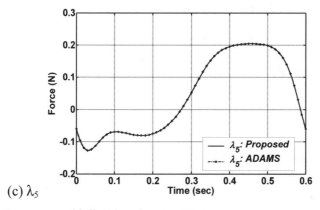

(c) λ_5

Fig. 3.20. Lagrange multipliers in subsystem II

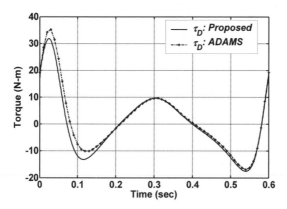

Fig. 3.21. Driving torque

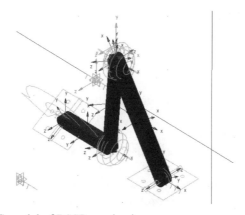

Fig. 3.22. ADAMS model of RSSR mechanism

3.5.3 Computation efficiency

The comparison of the theoretical order of computations and the CPU time taken by a Pentium IV computer is given in Table 3.5, where three different methods explained in Section 3.2 are used to solve the problem. The CPU time is estimated for the time duration of 0.6 sec in the time step of 0.001 sec. Note that the reduction in CPU time using system and subsystem approaches compared to traditional approach is not significantly less for the spatial problem. It is higher for the planar problem given in Table 3.3. The reason is that the time taken for transforming the vectors and matrices for the spatial problem is more in comparison to the planar problem. However, the proposed two-level recursive subsystem approach performs better for the spatial mechanism than the system approach. This is obvious from the reduction of the matrix sizes during calculation of the Lagrange multipliers and the driving torque.

Table 3.5. CPU time for 7R mechanism in Pentium IV(2.60 GHz)

Methods	Theoretical order of computations	CPU time in sec for dynamic analysis
Traditional (matrix size: 18×18)	$O(18^3/3)=O(1994)$	0.6560
System approach (matrix size:4×4)	$O(4^3/3+O(4×6)=O(45.33)$	0.609 $(7.16)^+$
Subsystem approach (matrix sizes: 3×3)	$O(3^3/3)+O(4×6)=O(33)$	0.5960 (9.15)

* For time duration of 0.6 sec (i.e., the time period of the crank) with steps of 0.001 sec. + Percentage savings in CPU time

3.6 Summary

In this chapter, a two-level recursive subsystem approach is discussed for the computation of constraint wrenches in planar and spatial mechanisms. The branches of the spanning tree of a closed-loop system are categorized into determinate and indeterminate subsystems. Equations of motion are derived systematically for each subsystem to find the Lagrange multipliers and driving torques/forces. The methodology is illustrated using two mechanisms, namely, the planar multiloop carpet scrapping mechanism, and a spatial RSSR mechanism.

4 Equimomental Systems

Dynamic quantities, e.g., shaking force, shaking moment, input-torque, etc., of a mechanical system depend on the mass and inertia of its each link, and the corresponding mass center location. These inertia properties can be represented more conveniently using the dynamically equivalent system of point-masses [110]. The dynamically equivalent system is also referred to as *equimomental system*. The concept is elaborated by Wenglarz et al. [111] and Haung [112]. In order to balance mechanisms, the concept of equimomental system of point-masses is introduced in this chapter.

4.1 Equimomental Systems for Planar Motion

A study on an equimomental system of rigidly connected point masses undergoing planar motion is presented in this section. A point mass is an idealized concept, and defined as a mass that is concentrated at a point. Two rigid systems are equimomental if their dynamic behaviors are identical, i.e., they have the same mass, the same center of mass, and the same inertia tensor with respect to a common point [110]. Considering a rigid body as shown in Fig. 4.1, and having mass m moving in the plane of diagram, a dynamically equivalent system of p point-masses fixed to the frame, OXY, is now sought. Its coordinate frame OXY fixed to the body at O. Assume that the mass of each point-mass is m_i, and located at the coordinates, (x_i, y_i). The system of p rigidly connected point-masses and the original rigid body are equimomental if they have: (a) the same mass; (b) the same center of mass; and (c) the same moment of inertia about an axis perpendicular to the plane of motion passing through O, i.e.,

$$\sum_{i=1}^{p} m_i = m \qquad (4.1)$$

$$\sum_{i=1}^{p} m_i x_i = m\bar{x} \tag{4.2}$$

$$\sum_{i=1}^{p} m_i y_i = m\bar{y} \tag{4.3}$$

$$\sum_{i=1}^{p} m_i (x_i^2 + y_i^2) = I^c + m(\bar{x}^2 + \bar{y}^2) \tag{4.4}$$

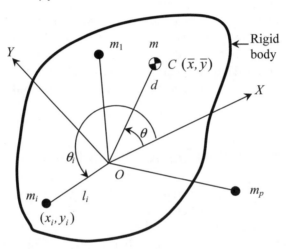

Fig. 4.1. A rigid body and its equimomental system of p point-masses

where I^c is the mass moment of inertia of the body about its mass center, C, whose coordinates are $(\bar{x}, \bar{y})$. Since each mass requires three parameters, (m_i, x_i, y_i), to identify it, a total of $3p$ parameters are necessary to completely define the equimomental system of p point-masses. However, there are four constraints, namely, Eqs. (4.1-4), that need to be satisfied. Hence, an infinite number of solutions exist for $p \geq 2$, as the resultant system of equations is underdeterminate, i.e., the number of unknowns is more than the equations [111]. If $p=1$, there is only one point-mass with three unknown parameters, which cannot satisfy all the four conditions, Eqs. (4.1-4) unless they are consistent. This is because the resulting system of equations is overdeterminate with more equations than the number of unknowns. Typically, such system of equations does not yield any solution unless the equations are consistent. As a consequence, an equimomental system of a rigid body moving in a plane cannot be represented using one point-mass, which is obvious from the fundamental knowledge of mechanics. Clearly, the minimum number of point-masses is then two giving six unknown parameters, of which two need to be assigned arbitrarily. If three point-masses are taken, five parameters are to be assigned arbitrarily. In general, $(3p-4)$ parameters need to be assigned arbitrarily so that the remaining four are determinate. Note here that it is not always possible to get all the point-masses positive. This, however, does not hindrance the process of representing the rigid body as long as the total mass and the moment of inertia about the mass center give positive values [113].

4.1.1 Two point-mass model

As explained in the previous section, an equimomental system of point-masses of a rigid body moving in a plane requires at least two point-masses. The representation of the body by the equimomental system of two point-masses is referred to as two point-mass model. Similarly, equimomental system of three point-masses is called three point-mass model, and so on. In this section, the two point-mass model is illustrated. Let such a two point-mass model for a rigid body is moving in the XY plane. The polar coordinates of the point-masses are (l_i, θ_i), for $i=1, 2$. Note that the point-masses are rigidly fixed in the local frame. The system of two point-masses is then equimomental to the rigid body if it satisfies the conditions given by Eqs. (4.1-4), i.e.,

$$m_1 + m_2 = m \tag{4.5}$$

$$m_1 l_1 \cos \theta_1 + m_2 l_2 \cos \theta_2 = md \cos \theta \tag{4.6}$$

$$m_1 l_1 \sin \theta_1 + m_2 l_2 \sin \theta_2 = md \sin \theta \tag{4.7}$$

$$m_1 l_1^2 + m_2 l_2^2 = I \tag{4.8}$$

where $d = \sqrt{\bar{x}^2 + \bar{y}^2}$ and $\theta \equiv \tan^{-1}(\bar{y}/\bar{x})$ are the polar coordinates of the mass center of the rigid body, C and I is the moment of inertia about origin of local frame, O.

For a rigid body of given mass, its centre location, and inertia, one can convert the body into an appropriate two point-masses using Eqs. (4.5-8). Since the four equations are nonlinear in six unknown parameters of point-masses, a judicious selection is required to choose for the arbitrary assigned parameters. Assuming $\theta_1 = 0$ and $\theta_2 = \pi/2$ [115], the four parameters of the two point-masses, namely, m_1, m_2, l_1, and l_2, are determined from Eqs. (4.5-8) as

$$m_1 = \frac{(\bar{x}^2 - \bar{y}^2 + md) \pm \sqrt{(\bar{x}^2 - \bar{y}^2 + md)^2 - 4mI\bar{x}^2}}{2I} \tag{4.9}$$

$$m_2 = m - m_1 \tag{4.10}$$

$$l_1 = \frac{\bar{x}}{m_1} \tag{4.11}$$

$$l_2 = \frac{\bar{y}}{m_2} \tag{4.12}$$

where $\bar{x} \equiv md \cos\theta$ and $\bar{y} \equiv md \sin\theta$. Hence, each point-mass has two solutions. If $\theta = 0$, i.e, the mass center of the rigid link lies on the X-axis, two sets of the solutions are as follows:

$$m_1 = \begin{cases} \dfrac{m^2 d^2}{I} \\ m \end{cases}; \; m_2 = \begin{cases} m - \dfrac{m^2 d^2}{I} \\ 0 \end{cases}; l_1 = \begin{cases} \dfrac{md}{m_1} \\ d \end{cases}; \text{ and } l_2 = \begin{cases} 0 \\ 0 \end{cases} \tag{4.13}$$

This way one can convert a given rigid body into an suitable two point-mass model.

4.1.2 Three point-mass model

In this section, the procedure of finding a three point-mass model is illustrated. Consider a three point-mass model for a rigid body moving in the XY plane. The polar coordinates of the point-masses are (l_i, θ_i), for $i=1, 2, 3$. Similar to two point-mass model, the three point-mass model would then be equimomental to the original rigid body if Eqs. (4.1-4) are satisfied, i.e.,

$$\sum_{i=1}^{3} m_i = m \tag{4.14}$$

$$\sum_{i=1}^{3} m_i l_i \cos\theta_i = md \cos\theta \tag{4.15}$$

$$\sum_{i=1}^{3} m_i l_i \sin\theta_i = md\sin\theta \tag{4.16}$$

$$\sum_{i=1}^{3} m_i l_i^2 = mk^2 \tag{4.17}$$

where $mk^2 \equiv I^c + md^2$ — k being the radius of gyration about the point, O. Note that there are nine unknown parameters of point-masses, namely, m_i, l_i, and θ_i, for $i=1, 2, 3$, and four equations, Eqs. (4.14-17). Hence, it is important to decide which five parameters should be chosen so that the remaining four become determinate. It is advisable to choose l_i and θ_i so that the dynamic equivalence conditions become linear in point masses. Assuming, $l_2 = l_3 = l_1$, and substituting them in Eq. (4.17) yields

$$\left(\sum_{i=1}^{3} m_i\right) l_1^2 = mk^2 \tag{4.18}$$

which gives $l_1 = \pm k$. Taking the positive value for l_1, which is physically possible, Eqs (4.14-16) are then written in a compact form as

$$\mathbf{Km} = \mathbf{b} \tag{4.19}$$

where the 3×3 matrix, $\mathbf{K}$, and the 3-vectors, $\mathbf{m}$ and $\mathbf{b}$, are as follows:

$$\mathbf{K} \equiv \begin{bmatrix} 1 & 1 & 1 \\ k\cos\theta_1 & k\cos\theta_2 & k\cos\theta_3 \\ k\sin\theta_1 & k\sin\theta_2 & k\sin\theta_3 \end{bmatrix} ; \mathbf{m} \equiv \begin{bmatrix} m_1 \\ m_2 \\ m_3 \end{bmatrix} ; \mathbf{b} \equiv \begin{bmatrix} m \\ md\cos\theta \\ md\sin\theta \end{bmatrix} \tag{4.20}$$

The magnitudes of three point-masses are then solved from Eq. (4.19) by assuming suitable values for θ_i, $i=1, 2, 3$. It is clear that the solution for $\mathbf{m}$ exists if $\det(\mathbf{K}) \neq 0$, i.e., $\theta_1 \neq \theta_2$, $\theta_1 \neq \theta_3$, and $\theta_2 \neq \theta_3$. It means that any two point masses should not lie on the same radial line emanating from the origin, O. The vector $\mathbf{m}$ is obtained as

$$\mathbf{m} = \mathbf{K}^{-1}\mathbf{b} \tag{4.21}$$

where $\mathbf{K}^{-1}$ is evaluated as

$$\mathbf{K}^{-1} \equiv \frac{k}{\det(\mathbf{K})} \begin{bmatrix} k\sin(\theta_3 - \theta_2) & (\sin\theta_2 - \sin\theta_3) & (\cos\theta_3 - \cos\theta_2) \\ -k\sin(\theta_3 - \theta_1) & (\sin\theta_3 - \sin\theta_1) & (\cos\theta_1 - \cos\theta_3) \\ -k\sin(\theta_1 - \theta_2) & (\sin\theta_1 - \sin\theta_2) & (\cos\theta_2 - \cos\theta_1) \end{bmatrix} \quad (4.22)$$

in which, $\det(\mathbf{K}) \equiv k^2[\sin(\theta_3 - \theta_2) + \sin(\theta_2 - \theta_1) + \sin(\theta_1 - \theta_3)]$. It is evident from the solution, Eq. (4.21), that the sum of the point-masses is equal to the mass of the body for any values of angles except $\theta_1 \neq \theta_2, \theta_1 \neq \theta_3$, and $\theta_1 \neq \theta_3$. Note here that, there is a possibility that some point masses are negative. It does not hindrance the process of representing the rigid body as long as its mass, m, and inertia, I^c, are positive and real, as pointed out before subsection 4.1.1. As an example, if $\theta_1 = 0$, $\theta_2 = 2\pi/3$, and $\theta_3 = 4\pi/3$, the point masses are calculated as

$$m_1 = \frac{m}{3}\left(1 + \frac{2d\cos\theta}{k}\right) \quad (4.23)$$

$$m_2 = \frac{m}{3}\left(1 - \frac{d\cos\theta}{k} + \frac{\sqrt{3}d\sin\theta}{k}\right) \quad (4.24)$$

$$m_2 = \frac{m}{3}\left(1 - \frac{d\cos\theta}{k} - \frac{\sqrt{3}d\sin\theta}{k}\right) \quad (4.25)$$

which take simpler form if the origin, O, coincides with the mass center of the body, C, i.e., $d=0$. Substituting $d=0$ in Eqs. (4.23-25), one obtains $m_1 = m_2 = m_3 = m/3$. It means that the point masses of the body is distributed equally, and located on the circumference of a circle having radius k.

It is pointed out here that in mechanism analysis, links are often considered as one-dimensional, e.g., a straight rod, in which its diameter or width and thickness are very small in comparison to the length. Considering that the mass lying along the X-axis of the local frame, the dynamical equivalence conditions, Eqs. (4.1-4), reduce to

$$\sum_{i=1}^{p} m_i = m; \quad \sum_{i=1}^{p} m_i x_i = m\bar{x}; \quad \sum_{i=1}^{p} m_i x_i^2 = I^c + m\bar{x}^2 \quad (4.26\text{-}28)$$

It is evident from Eqs. (4.26-28) that a minimum of two point-masses is also required to represent an one-dimensional link, introducing a total of four variables, i.e., m_1, m_2, x_1, and x_2. Specifying any one of the variables, the other three variables can be found uniquely. It is pointed out here that, a common practice in the dynamics study of reciprocating engine is to replace the connecting rod by two point-masses, where the masses are placed at the ends of the connecting rod. This does not provide a true equivalent system unless the three equations, Eqs. (4.26-28), in two unknowns, m_1 and m_2, leading to an overdetermined system are consistent.

4.2 Equimomental Systems for Spatial Motion

Consider a three-dimensional motion by a rigid body shown in Fig. 4.2, whose mass is m, the position of the mass center is $(\bar{x}, \bar{y}, \bar{z})$, the moments of inertia are, I_{xx}, I_{yy}, I_{zz} , and the products of inertia are, I_{xy}, I_{yz}, I_{zx}. The coordinates for the mass center position and the inertias are referred in the body fixed frame, $OXYZ$ of Fig. 4.2

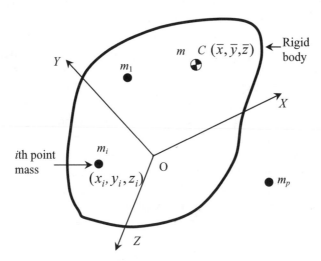

Fig. 4.2. Equimomental system of a rigid body in spatial motion

Let there be p point-masses, m_i, which are rigidly fixed to the frame at the positions (x_i, y_i, z_i), for $i=1, ..., p$. Now, two systems, i.e., the system of p point-masses and the rigid body, are dynamically equivalent if the following conditions are satisfied:

$$\sum_{i=1}^{p} m_i = m \qquad (4.29)$$

$$\sum_{i=1}^{p} m_i x_i = m\bar{x} \; ; \; \sum_{i=1}^{p} m_i y_i = m\bar{y} \; ; \; \sum_{i=1}^{p} m_i z_i = m\bar{z} \qquad (4.30\text{-}32)$$

$$\sum_{i=1}^{p} m_i x_i y_i = I_{xy} \; ; \; \sum_{i=1}^{p} m_i y_i z_i = I_{yz} \; ; \; \sum_{i=1}^{p} m_i z_i x_i = I_{zx} \qquad (4.33\text{-}35)$$

$$\sum_{i=1}^{p} m_i (y_i^2 + z_i^2) = I_{xx} \; ; \; \sum_{i=1}^{p} m_i (z_i^2 + x_i^2) = I_{yy} \; ; \; \sum_{i=1}^{p} m_i (x_i^2 + y_i^2) = I_{zz} \qquad (4.36\text{-}38)$$

Equation (4.29) ensures that the total mass of the equimomental system of point-masses is equal to the mass of the rigid body. Equations (4.30-32) satisfy the conditions of the mass center location, whereas Eqs. (4.33-38) ensure the same inertia tensor for the equivalent point mass system and the rigid body about point O. Since each point-mass is identified with four parameters, namely, m_i, x_i, y_i, z_i, the set of p point-masses requires $4p$ parameters that must satisfy ten conditions given by Eqs. (4.29-38). Hence, the smallest positive integer p that will be required to satisfy the ten constraints is three, as $4p \geq 10$. Note, however, that three points determine a plane. Moreover, in a planar system of mass distribution the moment of inertia about any axis perpendicular to the plane is the sum of the moments of inertia about any two perpendicular axes in the plane, which is obviously not a general property of an arbitrary body moving in the 3-dimesional Cartesian space [111]. As a result, three point masses are not sufficient to describe a rigid body motion in space, and a minimum of four point-masses are required to determine the equimomental system of point-masses [110].

Since Eqs. (4.30-38) are nonlinear in the variables, m_i, x_i, y_i, and z_i, the equations can be made linear in m_i if the coordinates, x_i, y_i, z_i, are specified for all the points. The solution for m_i is then depends upon the number of point-masses. For example, if $p = 4$, and x_i, y_i, z_i, for $i=1, \ldots, 4$, are specified, then four unknowns remain, namely, m_i for $i=1, \ldots, 4$, which cannot be solved from ten Eqs. (4.29-38) unless they are consistent. In order to obtain a determinate set of Eqs. (4.29-38) for the specified coordinates of all the point-masses, p must be ten, i.e., $p=10$. Alternatively, linearization of the equimomental conditions can be done by rearranging the moment of inertia conditions, Eqs. (4.36-38), as

$$\sum_{i=1}^{p} m_i x_i^2 = (-I_{xx} + I_{yy} + I_{zz})/2 \qquad (4.39)$$

$$\sum_{i=1}^{p} m_i y_i^2 = (I_{xx} - I_{yy} + I_{zz})/2 \qquad (4.40)$$

$$\sum_{i=1}^{p} m_i z_i^2 = (I_{xx} + I_{yy} - I_{zz})/2 \qquad (4.41)$$

Assuming the absolute values of the coordinates, x_i, y_i, and z_i, as h_x, h_y, and h_z, respectively, for all the p point-masses, and substituting them in Eqs. (4.39-41) yields

$$h_x^2 = \frac{-I_{xx} + I_{yy} + I_{zz}}{2m} \qquad (4.42)$$

$$h_y^2 = \frac{I_{xx} - I_{yy} + I_{zz}}{2m} \qquad (4.43)$$

$$h_z^2 = \frac{I_{xx} + I_{yy} - I_{zz}}{2m} \qquad (4.45)$$

where the condition of Eq. (4.29) is used, and the moments of inertia, I_{xx}, I_{yy}, and I_{zz}, are such that the sum of any two of them is always greater than the third one [110], i.e.,

$$(-I_{xx} + I_{yy} + I_{zz}) > 0 \; ; (I_{xx} - I_{yy} + I_{zz}) > 0 \; ; \text{and} \, (I_{xx} + I_{yy} - I_{zz}) > 0 \quad (4.46)$$

Therefore, h_x, h_y, and h_z never have imaginary values. Knowing h_x, h_y, and h_z from Eqs. (4.42-45), the coordinates of all the point-masses are obtained as follows:

$$x_i = \pm h_x, \quad y_i = \pm h_y, \quad z_i = \pm h_z, \text{ for } i=1, ..., p$$

The remaining unknowns, namely, m_1, ..., m_p, can then be determined uniquely from the remaining seven linear algebraic Eqs. (4.29-35) if $p=7$. Hence, a set of seven point-masses is proposed here to represent a rigid body moving in a space. For any other set of point masses Eqs. (4.29-35) are either overdeterminate or underdeterminate. Moreover, to avoid the co-incidence of two or more points, they must be placed at unique locations. This is only possible if all the points have unique coordinates. Since the positive and negative values of h_x, h_y, and h_z represent Cartesian coordinates of the point-masses, they form a rectangular parallelepiped whose center is at the origin, O, and the sides are $2h_x$, $2h_y$, and $2h_z$, as shown in Fig. 4.3. The coordinates of the point-masses are also indicated in Fig. 4.3.

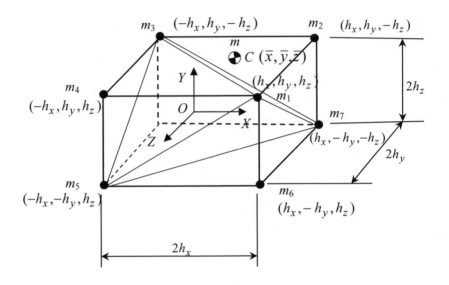

Fig. 4.3. Equimomental system of seven point-masses

Upon substitution of the coordinates of the point-masses shown in Fig. 4.3, the dynamically equivalent conditions, Eqs. (4.29-35), are obtained as

$$\sum_{i=1}^{7} m_i = m \tag{4.47}$$

$$[(m_1 + m_2 + m_6 + m_7) - (m_3 + m_4 + m_5)]h_x = m\bar{x} \tag{4.48}$$

$$[(m_1 + m_2 + m_3 + m_4) - (m_5 + m_6 + m_7)]h_y = m\bar{y} \tag{4.49}$$

$$[(m_1 + m_4 + m_5 + m_6) - (m_2 + m_3 + m_7)]h_z = m\bar{z} \tag{4.50}$$

$$[(m_1 + m_2 + m_5) - (m_3 + m_4 + m_6 + m_7)]h_x h_y = I_{xy} \tag{4.51}$$

$$[(m_1 + m_4 + m_7) - (m_2 + m_3 + m_5 + m_6)]h_y h_z = I_{yz} \tag{4.52}$$

$$[(m_1 + m_3 + m_6) - (m_2 + m_4 + m_5 + m_7)]h_z h_x = I_{zx} \tag{4.53}$$

Equations (4.47-53) are now written in a compact form as

$$\mathbf{Km} = \mathbf{b} \tag{4.54}$$

where the vector of unknowns, **m**, is solved as

$$\mathbf{m} = \mathbf{K}^{-1}\mathbf{b} \tag{4.55}$$

in which the 7×7 matrix, **K**, and $\mathbf{K}^{-1}$ are as follows:

$$\mathbf{K} \equiv \begin{bmatrix} 1 & 1 & 1 & 1 & 1 & 1 & 1 \\ 1 & 1 & -1 & -1 & -1 & 1 & 1 \\ 1 & 1 & 1 & 1 & -1 & -1 & -1 \\ 1 & -1 & -1 & 1 & 1 & 1 & -1 \\ 1 & 1 & -1 & -1 & 1 & -1 & -1 \\ 1 & -1 & -1 & 1 & -1 & -1 & 1 \\ 1 & -1 & 1 & -1 & -1 & 1 & -1 \end{bmatrix}; \text{ and}$$

$$\mathbf{K}^{-1} = \begin{bmatrix} 0.25 & 0 & 0 & 0 & 0.25 & 0.25 & 0.25 \\ 0 & 0.25 & 0.25 & 0 & 0 & -0.25 & -0.25 \\ 0.25 & -0.25 & 0 & -0.25 & 0 & 0 & 0.25 \\ 0 & 0 & 0.25 & 0.25 & -0.25 & 0 & -0.25 \\ 0.25 & -0.25 & -0.25 & -0 & 0.25 & 0 & 0 \\ 0 & 0.25 & 0 & 0.25 & -0.25 & -0.25 & 0 \\ 0.25 & 0 & -0.25 & -0.25 & 0 & 0.25 & 0 \end{bmatrix}$$

whereas the 7-vectors, **m** and **b**, are given by

$$\mathbf{m} \equiv \begin{bmatrix} m_1 & m_2 & m_3 & m_4 & m_5 & m_6 & m_7 \end{bmatrix}^T \text{ and}$$

$$\mathbf{b} \equiv \begin{bmatrix} m & \dfrac{m\bar{x}}{h_x} & \dfrac{m\bar{y}}{h_y} & \dfrac{m\bar{z}}{h_z} & \dfrac{I_{xy}}{h_x h_y} & \dfrac{I_{yz}}{h_y h_z} & \dfrac{I_{zx}}{h_z h_x} \end{bmatrix}^T$$

If the origin of the frame, O, coincides with the mass center, C, and the axes of the frame are along the principal axes, i.e., $\bar{x} = \bar{y} = \bar{z} = 0$ and $I_{xy} = I_{yz} = I_{zx} = 0$, it is evident from Eq. (4.55) that $m_1 = m_3 = m_5 = m_7 = 0.25\,m$, and $m_2 = m_4 = m_6 = 0$, irrespective of the values of h_x, h_y, and h_z. Therefore, the four point masses, m_1, m_3, m_5, and m_7, with equal masses are found. Such four point-mass model was used in [110-112] to represent a rigid body and used in [118-119] for the optimization of mass distribution of links. These point-masses form a tetrahedron, and each point lies at the vertices of the tetrahedron shown in Fig. 4.3. This tetrahedron is regarded as inscribed within a sphere of radius $\sqrt{h_x^2 + h_y^2 + h_z^2}$.

In case, one defines a new coordinate system (ξ, η, ζ) such that the coordinates of the point-masses are given as $x_i = h_x \xi_i$; $y_i = h_y \eta_i$; and $z_i = h_z \zeta_i$, for i=1, ...7, then the dynamically equivalent conditions of Eqs. (4.29-38) remain same except the terms, h_x, h_y, h_z, h_x^2, h_y^2, h_z^2, are replaced by unity. Assuming the new coordinate system (ξ, η, ζ) at the mass center and along the principal axes of the body, an equimomental system is obtained in the new coordinate system. This equimomental system is made of four point-masses having equal mass lying at the vertices of the regular tetrahedron inscribed within a sphere of radius √3. The results are conforming with those reported by Routh [110], i.e., "*Four particles of equal mass can always be found which are equimomental to any given solid body.*" However, in [110], no proof or justification was provided, which is reported in [134] for the first time. Also it is confirmed that the equimomental system form a tetrahedron with four equal particles. Later, Haug [112] formulated the problem to find the location of the four equal point-masses, but did not give any proof of its existence. Note that in the balancing of mechanisms, one needs the mass distribution of the links. Therefore, the determination of only the location of the point-masses having equal masses is not useful. Rather, the set of seven point-masses just described is quite general and more useful to formulate an optimization problem for the balancing of mechanisms, where the point-masses are the design variables.

5 Balancing of Planar Mechanisms

The dynamic balancing of mechanisms is an old problem in order to (1) reduce the amplitude of vibration of the frame on which the mechanism is mounted due to the shaking force and shaking moment; and (2) smoothen highly fluctuating driving torque/force needed to obtain nearly constant drive speed. Since any vibration leads to noise, wear, fatigue, etc., in the mechanism, its reduction improves several aspects of mechanical design as well. However, the problem is faced with new challenges, namely, the balancing of shaking force, shaking moment, and input-torque fluctuations together. The shaking force can be eliminated completely by attaching counterweights to the moving links. However, this increases overall mass and inertia of the mechanism. As a result, shaking moment, required driving torque, and reactions at the joints increase. Hence, to improve the overall performance, it is required to trade-off between all the competing dynamic quantities, namely, the shaking force, shaking moment, driving torque/force, bearing reactions, etc. This means that the balancing problem should be treated as an optimization problem, whose formulation needs the following:

1. an efficient dynamic algorithm to compute the dynamic quantities;

2. identification of the design variables, and the formulation of the constraints on them that define the design space of the feasible solutions; and

3. an objective function which is to be used as an index of merit for the dynamic performance of a mechanism at hand.

In this chapter, the balancing problem of planar mechanisms is formulated as a general mathematical optimization problem. The shaking force, shaking moment, driving torque/force, etc., depend on the mass and inertia of each link, and its mass center location for given dimensions and motion [113]. Hence, it is required to optimally distribute link masses for the dynamic balancing. In order to achieve this, each link is treated as a dynamically equivalent system or equimomental system of point masses. The parameters of the equimomental point-masses are then used as the design variables. Accordingly, the equations of motion proposed in Chapters 2 and 3 are modified for the optimization purpose.

5.1 Balancing of Shaking Force and Shaking Moment

Balancing of shaking force and shaking moment in the mechanisms is important in order to improve their dynamic performances and fatigue life by reducing vibration, noise and wear. Several methods are developed to eliminate the shaking force and shaking moment in planar mechanisms [81-102]. The methods to completely eliminate the shaking force are generally based on two principles: (1) making the total potential energy of a mechanism constant [81, 82], and (2) making the total mass center of a mechanism stationary [83-85]. Studies based on potential energy use elastic elements like springs to balance the force. On the other hand, the methods based on making total mass center stationary use mass redistribution/counterweights. Different techniques are used for tracing and making the total mass center stationary. For example, the method of *principal vectors* [84] describes the position of the mass center by a series of vectors that are directed along the links. These vectors trace the mass center of the mechanism at hand, and the conditions are derived to make the system mass center stationary. A more referred method in the literature is the method of *linearly independent vectors* [85] where the stationary condition was achieved by redistributing the link masses in such a manner that the coefficients of the time-dependent terms of the equations describing the total center of the mass trajectory vanish. Kosav [86] presented a general method using ordinary vector algebra instead of the complex number representation of the vectors [85] for full force balance of the planar linkages. One of the attractive features of a force-balanced linkage is that the shaking force vanishes, and the shaking moment reduces to a pure torque which is independent of reference point. However, only shaking force balancing is not effective in the balancing of mechanisms, as (1) it mostly increases the total mass of the mechanism, (2) it needs some arrangement like counterweights that increase the total mass, and (3) it increases the other dynamic characteristics, like shaking moment, driving torque, and bearing reactions. The influence of the complete shaking force balancing is thoughtfully investigated by Lowen et al. [87] on the bearing reactions, input-torque, and shaking moment for a family of crank-rocker four-bar linkages. This study shows that these dynamic quantities increase, and in some cases their values rise up to five-times.

Several authors attempted to treat the balancing problem as a complete shaking force and shaking moment balancing. Elliot et al. [88] developed a theory to balance torque, shaking force, and shaking moment by extending the method of linearly independent vectors. Complete moment balancing is also achieved by a cam-actuated oscillating counterweight [89], inertia counterweight [90, 91], physical pendulum [92], and geared counterweights [93-95]. More information on complete shaking moment balancing can be obtained in a critical review by Lowen et al. [83, 129], Kosav [96], and Arakelian and Smith [97]. Practically, these methods not only increase the mass of the system but also increase its complexity.

An alternate way to reduce the shaking force and shaking moment along with other dynamic quantities such as input-torque, bearing reactions, etc., is to optimize all the dynamic quantities. Since shaking moment reduces to a pure torque in a force-balanced linkage, many researchers used the fact to develop their theory of shaking moment optimization. Berkof and Lowen [98] proposed an optimization method to minimize the root-mean-square value of the shaking moment in a fully force-balanced in-line four-bar linkage whose input link is rotating at a constant speed. As an extension of this method, Carson and Stephens [99] highlighted the need to consider feasibility limits of the link parameters. A different approach for the optimization of shaking moment in a force-balanced four-bar linkage is proposed by Hains [100]. Using the principle of the independence of the static balancing properties of a linkage from the axis of rotation of the counterweights, partial shaking moment balancing is suggested by Arakelian and Dahan [101]. The principle of momentum conservation is also used by Wiederich and Roth [102] to reduce the shaking moment in a fully force-balanced four-bar linkage.

5.1.1 Equimomental system in optimization

Dynamic quantities, e.g., shaking force, shaking moment, input-torque, etc., depend on the mass and inertia of each link, and its mass center location. These inertia properties of a mechanism can be represented more conveniently using the dynamically equivalent system of point-masses [110], as explained in chapter 4. The dynamically equivalent system is also referred to as *equimomental system*. The concept is elaborated by Wenglarz et al. [111] and Haung [112]. Using the concept of equimomental system, Sherwood and Hokey [113] presented the optimization of mass distribution in mechanisms. Hockey [114] discussed the input-torque fluctuations of mechanisms subject to external loads by means of properly distributing the link masses. Using the two point-mass model, momentum balancing of four-bar linkages was presented in [102]. Optimum balancing of combined shaking force, shaking moment, and torque fluctuations in high speed linkages was reported in Lee and Cheng [115], and Qi and Pennestri [116], where a two point-mass model was used. The concept can also be applied for the kinematic and dynamic analyses of mechanisms [77,117].

Simultaneous minimization of shaking force, shaking moment, and other quantities using the dynamical equivalent system of point-masses and optimum mass distribution has been attempted in [118, 119]. However, the results do not show significant improvement in the performances.

5.2 Balancing Problem Formulation

The problem of mechanism balancing is formulated here as an optimization problem. In order to identify the design variables and the associated constraints, a set of equimomental point-masses is defined for each link of a mechanism at hand. To calculate the shaking force and shaking moment dynamic equations of motion in the minimal set are derived in the parameters of the point-masses. These parameters are then treated as design variables to redistribute the link masses to improve the dynamic performance of the mechanism.

5.2.1 Equations of motion

Referring to the ith rigid link, Fig. 5.1(a), of a planar mechanism, the location of its mass center, C_i, is defined by the vector, $\mathbf{d}_i$ at an angle, θ_i, from the axis O_iX_i of the local frame, $O_iX_iY_i$, fixed to the link. The axis O_iX_i is set along the vector from O_i to O_{i+1}, that is at an angle α_i from the axis, OX, of the fixed inertial frame, OXY, Fig. 5.1(a). The points, O_i and O_{i+1}, on the link are chosen as the points where the ith link is coupled to its neighboring links, whereas link's mass and the mass moment of inertia about O_i are m_i and I_i, respectively. A system of p point-masses, which is equimomental to the ith link, is shown in Fig. 5.1(b). The point-masses, m_{ij}, for $j=1, \ldots, p$, are fixed in the local frame, $O_iX_iY_i$, and their distances from the origin, O_i, are l_{ij}. The angles, θ_{ij}, are defined between the line joining the point masses from O_i, and the axis, O_iX_i. Note that the minimum number of point-masses required to represent the ith rigid link in plane motion is two, i.e., $p \geq 2$. In this section, all the vectors are represented in the fixed frame, OXY, unless stated otherwise.

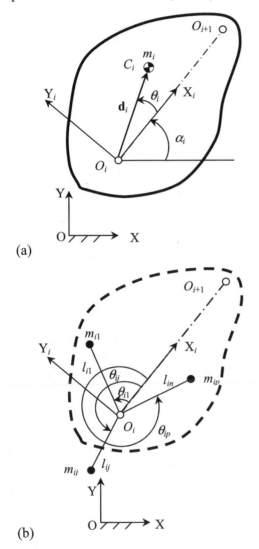

(a)

(b)

Fig. 5.1. Parameters for rigid link and its equimomental system (a) The ith rigid link (b) Equimomental system of the ith link

If the p point-masses is equimomental to the ith link, then they must satisfy the conditions given by Eqs. (4.1)-(4.4) referred to the fixed frame, OXY, i.e.,

$$\sum_{j=1}^{p} m_{ij} = m_i \tag{5.1}$$

$$\sum_{j=1}^{p} m_{ij} \mathbf{l}_{ij} = m_i \mathbf{d}_i \tag{5.2}$$

$$\sum_{j=1}^{p} m_{ij} l_{ij}^2 = I_i \tag{5.3}$$

where   2-vectors, $\mathbf{l}_{ij} \equiv [l_{ij} \cos(\theta_{ij} + \alpha_i) \quad l_{ij} \sin(\theta_{ij} + \alpha_i)]^T$ and

$\mathbf{d}_i \equiv [d_i \cos(\theta_i + \alpha_i) \quad d_i \sin(\theta_i + \alpha_i)]^T$ denote the position of m_{ij} with respect to O_i and the position of m_i from O_i, respectively. The first subscript i denotes the link number, and the second one, i.e., $j=1, \ldots, p$, represents the point-masses corresponding to the ith link. The equations of motion are now derived in terms of the point-masses using two approaches. In the first approach, the dynamic equivalent conditions are applied to the mass matrices associated with the dynamic equations of motion for a rigid body derived in Chapter 2. The second approach is to write the equations of motion of the point masses from the fundamentals, as done in next subsection. Here the first approach is shown, where the Newton-Euler (NE) equations of motion for the ith rigid link moving in a plane are reproduced from Eq. (2.37-39) as

$$\mathbf{M}_i \dot{\mathbf{t}}_i + \mathbf{C}_i \mathbf{t}_i = \mathbf{w}_i \tag{5.4}$$

where the 3-vectors, $\mathbf{t}_i, \dot{\mathbf{t}}_i$ and $\mathbf{w}_i$, are defined as the twist, twist-rate and wrench of the ith link with respect to the origin, O_i, i.e.,

$$\mathbf{t}_i \equiv \begin{bmatrix} \omega_i \\ \mathbf{v}_i \end{bmatrix}; \dot{\mathbf{t}}_i \equiv \begin{bmatrix} \dot{\omega}_i \\ \dot{\mathbf{v}}_i \end{bmatrix} \text{ and } \mathbf{w}_i \equiv \begin{bmatrix} n_i \\ \mathbf{f}_i \end{bmatrix} \tag{5.5}$$

in which ω_i and $\mathbf{v}_i$ are the scalar angular velocity about the axis perpendicular to the plane of motion, and the 2-vector of linear velocity of point O_i of the ith link respectively. Accordingly, $\dot{\omega}_i$ and $\dot{\mathbf{v}}_i$ are the time derivatives of ω_i and $\mathbf{v}_i$, respectively. Also, the scalar, n_i, and the 2-vector, $\mathbf{f}_i$, are the resultant moment about O_i and the resultant force at O_i, respectively. Moreover, the 3×3 matrices, $\mathbf{M}_i$ and $\mathbf{C}_i$ are reproduced from Chapter 2 as

$$\mathbf{M}_i \equiv \begin{bmatrix} I_i & -m_i \mathbf{d}_i^T \overline{\mathbf{E}} \\ m_i \overline{\mathbf{E}} \mathbf{d}_i & m_i \mathbf{1} \end{bmatrix} \text{ and } \mathbf{C}_i \equiv \begin{bmatrix} 0 & \mathbf{0}^T \\ -m_i \omega_i \mathbf{d}_i & \mathbf{O} \end{bmatrix} \tag{5.6}$$

where $\mathbf{1}$ and $\mathbf{O}$ are the 2×2 identity and zero matrices, respectively, and $\mathbf{0}$ is the 2- vector of zeros, and the 2×2 matrix, $\overline{\mathbf{E}}$, is defined by

$$\overline{\mathbf{E}} \equiv \begin{bmatrix} 0 & -1 \\ 1 & 0 \end{bmatrix}$$

Upon substitution of the expressions for the scalar, I_i, and the 2-vector, $m_i \mathbf{d}_i$, from Eqs. (5.1-3), the 3×3 matrices, $\mathbf{M}_i$ and $\mathbf{C}_i$ of Eq. (5.6) are obtained as

$$\mathbf{M}_i \equiv \begin{bmatrix} \displaystyle\sum_j m_{ij} l_{ij}^2 & -\displaystyle\sum_j m_{ij} l_{ij} \sin(\theta_{ij} + \alpha_i) & \displaystyle\sum_j m_{ij} l_{ij} \cos(\theta_{ij} + \alpha_i) \\ -\displaystyle\sum_j m_{ij} l_{ij} \cos(\theta_{ij} + \alpha_i) & \displaystyle\sum_j m_{ij} & 0 \\ \displaystyle\sum_j m_{ij} l_{ij} \sin(\theta_{ij} + \alpha_i) & 0 & \displaystyle\sum_j m_{ij} \end{bmatrix} \tag{5.7}$$

$$\mathbf{C}_i \equiv \begin{bmatrix} 0 & 0 & 0 \\ -\omega_i \displaystyle\sum_j m_{ij} l_{ij} \cos(\theta_{ij} + \alpha_i) & 0 & 0 \\ -\omega_i \displaystyle\sum_j m_{ij} l_{ij} \sin(\theta_{ij} + \alpha_i) & 0 & 0 \end{bmatrix}$$

Equations (5.4) and (5.7) are the equations of motion for the ith link in terms of its $3p$ point-mass parameters, namely, m_{ij}, θ_{ij}, l_{ij}, for $j=1, \dots, p$. Now, all or some of the point-mass parameters can be used as design variables based on their influence on the objective function of an optimization problem. For example, as evident from Eq. (5.3), the angles, θ_{ij}, do not influence the moment of inertia, however, the distances, l_{ij}, do. Hence, the angles, θ_{ij}, are excluded from the set of design variables.

In some research papers, namely, by Wiederrich and Roth [102], Lee and Cheng [115], and Qi and Pennestri [116], two point-mass model was considered to represent the mass and inertia of a link. They assumed, $\theta_{i1} = 0$ and $\theta_{i2} = \pi/2$, amongst the six parameters m_{i1}, m_{i2}, θ_{i1}, θ_{i2}, l_{i1}, and l_{i2}. The remaining parameters were then considered as design variables, and used for the optimization of four-bar mechanisms. Here three-point mass model as proposed in chapter 4 is used, where the following five parameters are assigned arbitrarily:

$$\theta_{i1} = 0 \; ; \theta_{i2} = 2\pi/3 \; ; \; \theta_{i3} = 4\pi/3 \; ; \text{and } l_{i2} = l_{i3} = l_{i1} \tag{5.8}$$

The other four parameters, namely, m_{i1}, m_{i2}, m_{i3}, and l_{i1}, are then treated as the design variables for each link.

The equations of motion in terms of the point-mass parameters, Eq. (4.22), for n moving links of a mechanism, i.e., i=1, ..., n, are then expressed as

$$\mathbf{M\dot{t} + Ct = w} \tag{5.9}$$

where all the vectors and matrices are defined in Chapter 2. Equation (5.9) is essentially $3n$ scalar equations in $3n$ parameters of point-masses. These equations of motion are the unconstrained equations of motion similar to the unconstrained NE equations of motion of n uncoupled rigid bodies given en derived in Chapter 2. Next, to compute the constraint wrenches from Eq. (5.9) that are necessary to find the shaking force and shaking moment the proposed two-level recursive method is used due to its obvious advantages highlighted in Section 3.2.

5.2.2 Equations of motion for a point-mass system

Since a rigid body is represented in subsection 5.2.1 as a set of point-masses, one should be able to derive the equations of motion, Eqs. (5.4) and (5.7), are only from the Newton's equations of linear motion. To validate the same, a set of three rigidly connected point-masses shown in Fig. 5.2 is considered. This set has three degree of freedom (DOF) and three independent constraints, namely, the distances between the point masses remain same. Hence, the given system of point-masses can be represented by three independent equations of motion, which is of course true as the the point-mass system represents a rigid body motion moving on a plane. The Newton's equations of motion for the three points are given by

$$m_i \ddot{x}_i = f_{ix} \tag{5.10}$$

$$m_i \ddot{y}_i = f_{iy} \tag{5.11}$$

for i=1, 2, 3, where f_{ix} and f_{iy} are the components of the resultant force acting on the ith point-mass along X and Y axes, respectively.

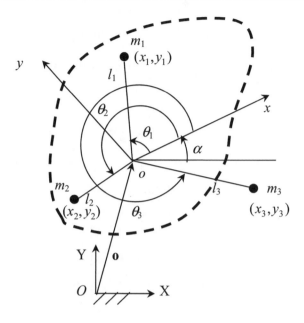

Fig. 5.2. Three rigidly connected point masses

To apply the concept of the natural orthogonal complement (NOC) in deriving the equations of motion of the point-mass system, the Cartesian coordinates of each point mass, (x_i, y_i), are first expressed from Fig. 5.2 as

$$x_i = o_x + l_i \cos(\theta_i + \alpha) \qquad (5.12)$$

$$y_i = o_y + l_i \sin(\theta_i + \alpha) \qquad (5.13)$$

in which o_x and o_y are the components of vector **o** that denotes the origin o of frame oxy with respect to the origin of the inertia frame, O. Differentiating Eqs. (5.12-13) once and twice with respect to time yields

$$\dot{x}_i = \dot{o}_x - l_i \sin(\theta_i + \alpha)\dot{\alpha} \qquad (5.14)$$

$$\dot{y}_i = \dot{o}_y + l_i \cos(\theta_i + \alpha)\dot{\alpha} \qquad (5.15)$$

and

$$\ddot{x}_i = \ddot{o}_x - l_i \cos(\theta_i + \alpha)\dot{\alpha}^2 - l_i \sin(\theta_i + \alpha)\ddot{\alpha} \qquad (5.16)$$

$$\ddot{y}_i = \ddot{o}_y - l_i \sin(\theta_i + \alpha)\dot{\alpha}^2 + l_i \cos(\theta_i + \alpha)\ddot{\alpha} \tag{5.17}$$

Using Eqs. (5.14-15), the generalized twist of the system of three point-masses is then given by

$$\mathbf{t} = \mathbf{N}\dot{\mathbf{q}} \tag{5.18}$$

where $\mathbf{t} \equiv [\mathbf{t}_1^T, \cdots, \mathbf{t}_3^T]^T$ in which $\mathbf{t}_i \equiv [\dot{x}_i, \dot{y}_i]^T$, $\dot{\mathbf{q}} \equiv [\dot{\alpha}, \dot{o}_x, \dot{o}_y]^T$, and $\mathbf{N}$ is the 6×3 natural orthogonal complement (NOC) matrix [68] given by

$$\mathbf{N} \equiv \begin{bmatrix} -l_1 \sin(\theta_1 + \alpha) & 1 & 0 \\ l_1 \cos(\theta_1 + \alpha) & 0 & 1 \\ -l_2 \sin(\theta_2 + \alpha) & 1 & 0 \\ l_2 \cos(\theta_2 + \alpha) & 0 & 1 \\ -l_3 \sin(\theta_3 + \alpha) & 1 & 0 \\ l_3 \cos(\theta_3 + \alpha) & 0 & 1 \end{bmatrix} \tag{5.19}$$

Now, the Newton equations of motion, Eqs. (5.10-11), for the *i*th point-mass are represented as

$$\mathbf{M}_i \dot{\mathbf{t}}_i = \mathbf{w}_i \tag{5.20}$$

where the 2×2 mass matrix, $\mathbf{M}_i$, and the 2-vector of wrench, $\mathbf{w}_i$, are defined by

$$\mathbf{M}_i \equiv \begin{bmatrix} m_i & 0 \\ 0 & m_i \end{bmatrix}; \mathbf{w}_i \equiv \begin{bmatrix} f_{ix} \\ f_{iy} \end{bmatrix}$$

In Eq. (5.20), the twist rate, $\dot{\mathbf{t}}_i$, is expressed from Eqs. (5.16-17) as

$$\dot{\mathbf{t}}_i \equiv \begin{bmatrix} \ddot{x}_i \\ \ddot{y}_i \end{bmatrix} = \begin{bmatrix} \ddot{o}_x - l_i \cos(\theta_i + \alpha)\dot{\alpha}^2 - l_i \sin(\theta_i + \alpha)\ddot{\alpha} \\ \ddot{o}_y - l_i \sin(\theta_i + \alpha)\dot{\alpha}^2 + l_i \cos(\theta_i + \alpha)\ddot{\alpha} \end{bmatrix} \tag{5.21}$$

The equations of motion, Eq. (5.20), for the three point-masses are then put together as

$$\mathbf{M}\dot{\mathbf{t}} = \mathbf{w} \tag{5.22}$$

where the 6×6 generalized mass matrix, $\mathbf{M} \equiv diag[\mathbf{M}_1, \cdots, \mathbf{M}_3]$, and the 6-vector of generalized wrench, $\mathbf{w} \equiv [\mathbf{w}_1^T, \cdots, \mathbf{w}_3^T]^T$. Pre-multiplying the transpose of the natural orthogonal complement, $\mathbf{N}$, to Eq. (5.22) gives the minimal set of constrained equations of motion, namely,

$$\mathbf{N}^T \mathbf{M} \dot{\mathbf{t}} = \mathbf{N}^T \mathbf{w} \qquad (5.23)$$

Upon substitution of the expressions of $\mathbf{N}$ and $\dot{\mathbf{t}}_i \equiv \begin{bmatrix} \mathbf{t}_1^T & \mathbf{t}_2^T & \mathbf{t}_3^T \end{bmatrix}^T$ from Eqs. (5.19) and (5.21), respectively, the equations of motion, Eq. (5.23), are obtained as

$$
\begin{bmatrix}
\sum_i m_i l_i^2 & -\sum_i m_i l_i \sin(\theta_i + \alpha) & \sum_i m_i l_i \cos(\theta_i + \alpha) \\
-\sum_i m_i l_i \cos(\theta_i + \alpha) & \sum_i m_i & 0 \\
\sum_i m_i l_i \sin(\theta_i + \alpha) & 0 & \sum_i m_i
\end{bmatrix}
\begin{bmatrix} \ddot{\alpha} \\ \ddot{o}_x \\ \ddot{o}_y \end{bmatrix}
\qquad (5.24)
$$

$$
+
\begin{bmatrix}
0 & 0 & 0 \\
-\dot{\alpha}\sum_i m_i l_i \cos(\theta_i + \alpha) & 0 & 0 \\
-\dot{\alpha}\sum_i m_i l_i \sin(\theta_i + \alpha) & 0 & 0
\end{bmatrix}
\begin{bmatrix} \dot{\alpha} \\ \dot{o}_x \\ \dot{o}_y \end{bmatrix}
$$

$$
=
\begin{bmatrix}
-\sum_i m_i l_i f_{ix} \sin(\theta_i + \alpha) + \sum_i m_i l_i f_{iy} \cos(\theta_i + \alpha) \\
\sum_i f_{ix} \\
\sum_i f_{iy}
\end{bmatrix}
$$

In Eq. (5.24), the right hand side is the vector of generalized forces, $\mathbf{w}_i$ of Eq. (5.4), whereas the generalized coordinates, velocities, and accelerations are, $\mathbf{q} \equiv [\alpha \quad o_x \quad o_y]^T$, $\dot{\mathbf{q}} \equiv [\dot{\alpha} \quad \dot{o}_x \quad \dot{o}_y]^T$ and $\ddot{\mathbf{q}} \equiv [\ddot{\alpha} \quad \ddot{o}_x \quad \ddot{o}_y]^T$, respectively. The matrices associated with the vectors, $\ddot{\mathbf{q}}$ and $\dot{\mathbf{q}}$, are nothing but the 3×3 matrices, $\mathbf{M}_i$ and $\mathbf{C}_i$, of Eq. (5.7), respectively. Also, $\dot{\alpha}$ in Eq. (5.24) represents the angular velocity of the rigid body, and $\dot{o}_x$ and $\dot{o}_y$ are the components of the linear velocity of the rigid body which is represented by the three point-mass system. Hence, $\dot{\mathbf{q}}$ and $\ddot{\mathbf{q}}$ are nothing but the twist, $\mathbf{t}_i$, and twist rate $\dot{\mathbf{t}}_i$ of the ith rigid body, respectively, given in given (5.5).

5.2.3 Shaking force and shaking moment

Figure 5.3 shows n moving links in a multiloop mechanism where the fixed link, #0, is detached from the other links. The appropriate reaction forces and moments due to the fixed link are indicated on the moving links to maintain the dynamic equilibrium.

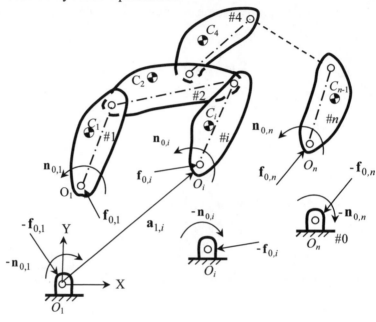

Fig. 5.3. A multiloop mechanism

The shaking force is now defined as the reaction of the vector sum of all the inertia forces of moving links associated with the mechanism, and the shaking moment is the reaction of the resultant of the inertia moment and the moment of the inertia forces [115]. By the above definitions, the shaking force and the shaking moment with respect to O_1, transmitted to the fixed link are given by

$$\mathbf{f}_{sh} = -\sum_{i=1}^{n} \mathbf{f}_i^*$$

(5.25)

$$n_{sh} = -\sum_{i=1}^{n} (n_i^* - \mathbf{a}_{1,i}^T \overline{\mathbf{E}} \mathbf{f}_i^*)$$

(5.26)

where n_i^* and $\mathbf{f}_i^*$ are the inertia moment and the 2-vector of inertia force, respectively, acting at and about the origin, O_i, of the ith link. Moreover, the 2-vector, $\mathbf{a}_{1,i}$, is defined from O_1 to the origin of the ith link, as shown in Fig. 5.3. Substituting the resultant force and moment in terms of the external force and moment, and the reactions due to the adjoining joints, the force and moment balance expressions for the ith link are written as

$$\mathbf{f}_i^* = \mathbf{f}_i^e + \sum_{k=0}^{n} \mathbf{f}_{k,i} \tag{5.27}$$

$$n_i^* = n_i^e + \sum_{k=0}^{n} (n_{k,i} - \mathbf{a}_{i,k}^T \overline{\mathbf{E}} \mathbf{f}_{k,i}) \tag{5.28}$$

where $\mathbf{f}_{k,i}$ and $n_{k,i}$ are the bearing reaction force and moment on the ith link by the kth link, respectively. Note that $\mathbf{f}_{k,i} = \mathbf{0}$ and $n_{k,i} = 0$ if kth link is not directly connected to the ith link. Furthermore, $\mathbf{f}_i^e$ and n_i^e are the external force and moment acting at and about the origin, O_i, respectively. Note that the origin for the ith link is defined at the joint where it is coupled with its previous body, whereas vector, $\mathbf{a}_{i,k}$, is defined from the origin of the ith link to the joint where the kth link is connected. Upon substitution of Eqs. (5.27-28) into Eqs. (5.25-26), the shaking force, and the shaking moment with respect to O_1 transmitted to the fixed link, #0, are obtained as

$$\mathbf{f}_{sh} = -\sum_{j=1}^{n_f} \mathbf{f}_{0,j} - \sum_{i=1}^{n} \mathbf{f}_i^e \tag{5.29}$$

$$n_{sh} = -\sum_{j=1}^{n_f} (n_{0,j} - \mathbf{a}_{1,j}^T \overline{\mathbf{E}} \mathbf{f}_{0,j}) - \sum_{i=1}^{n} (n_i^e - \mathbf{a}_{1,i}^T \overline{\mathbf{E}} \mathbf{f}_i^e) \tag{5.30}$$

where $\mathbf{f}_{0,j}$ represents the reaction force on the jth link by the fixed link, for $j=1, \ldots, n_f$ — n_f being the number of links connected to the fixed link. Hence, using Eqs. (5.29-30), the computation of the reactions at all the joints is not necessary to compute the shaking force and shaking moment Note that the dynamic quantities, e.g., the shaking force, shaking moment, and bearing reactions, have different units and magnitudes. In order to harmonize them, the force and moment are normalized as [125]:

$$\bar{f} = \left|\mathbf{f}\right| / (m_m^o a_m \omega_{in}^2) \tag{5.31}$$

$$\bar{n} = n / (m_m^o a_m^2 \omega_{in}^2) \tag{5.32}$$

where a_m and m_m^o are the length and mass of the reference link for the normalization, whereas ω_{in} is any input angular velocity. Superscript 'o' is used for those parameters of the original mechanism, which will be changing during the optimization.

5.2.4 Optimality criterion

There are many possible criteria by which the shaking force and Shaking moment transmitted to the fixed link of the mechanism can be minimized. For example, one criterion could be based on the root mean squares (RMS) values of the shaking force, shaking moment, and the required driving torque for a given motion, and/or the combination of these. Besides the RMS values, there are other ways to specify the dynamic quantities also, namely, by the maximum values, or by the amplitude of the specified harmonics, or by the amplitudes at certain point during the motion cycle. The choice of course depends on the requirements. Here, the RMS value is preferred over others as it gives equal emphasis on the results of every time instances, and every harmonic component.   The root mean square (RMS) values of the normalized shaking force, $\tilde{f}_{sh}$, and the normalized shaking moment, $\tilde{n}_{sh}$, at δ discrete positions of the mechanism are defined as

$$\tilde{f}_{sh} \equiv \sqrt{\sum \bar{f}_{sh}^2 / \delta} \; ; \text{ and } \tilde{n}_{sh} \equiv \sqrt{\sum \bar{n}_{sh}^2 / \delta} \tag{5.33}$$

where $\tilde{f}_{sh}$ and $\tilde{n}_{sh}$ are the RMS valuesof the normalized shaking force and the normalized shaking moment, respectively. Considering the RMS values, $\tilde{f}_{sh}$ and $\tilde{n}_{sh}$, an optimality criterion is proposed as

$$z = w_1 \tilde{f}_{sh} + w_2 \tilde{n}_{sh} \tag{5.34}$$

where w_1 and w_2 are the weighting factors whose values may vary depending on an application. For example, $w_1=1.0$ and $w_2=0$ if the objective is to minimize the shaking force only. The design variables and constraints depend upon whether the balancing is done through the redistribution of link masses or counterweighting the links.

5.2.5 Mass redistribution method

Consider a mechanism having n moving links, i.e., $i=1, \ldots, n$, and each link is modeled by a system of p equimomental point-masses, then the $3p$-vector of point-mass parameters for the ith link is defined as

$$\mathbf{x}_i \equiv [m_{i1} \quad \cdots \quad m_{ip} \quad l_{i1} \quad \cdots \quad l_{ip} \quad \theta_{i1} \quad \cdots \quad \theta_{ip}]^T \qquad (5.35)$$

Accordingly, the $3np$-vector of the point-mass parameters for the whole mechanism is given by

$$\mathbf{x} = [\mathbf{x}_i^T \quad \cdots \quad \mathbf{x}_n^T]^T \qquad (5.36)$$

If three point-mass model is used then the dimension of the vectors, $\mathbf{x}_i$ and $\mathbf{x}$, are 9 and $9n$, respectively. The dimensions of the vector, $\mathbf{x}_i$ or $\mathbf{x}$, can, however, be reduced, as explained in Subsection 5.2.1. For the three point-mass model, if five parameters per link are assigned arbitrarily according to Eq. (5.8), the remaining four parameters, namely, m_{i1}, m_{i2}, m_{i3}, and l_{i1}, per link can be treated as the *design variables (DV)*. Finally the $4n$-vector, $\mathbf{x}$, of the DVs using three point-mass model is defined as

$$\mathbf{x} \equiv [m_{11}, m_{12}, m_{13}, l_{11}, \cdots, m_{n1}, m_{n2}, m_{n3}, l_{n1}]^T \qquad (5.37)$$

The constraints on the DVs depend on the allowable minimum and maximum values of the DVs, say, mass and inertia, etc. The minimum mass, $m_{i,\min}$, of the ith link and its mass distribution can be decided by the strength of its material. Furthermore, the maximum mass, $m_{i,\max}$, can be taken into account according to what extent the shaking force and shaking moment are eliminated. Similarly, the limits on parameters, l_{i1}, can be determined based on the limiting values of the moment of inertia. The optimization problem is finally posed as

$$\text{Minimize } z(\mathbf{x}) = w_1 \tilde{f}_{sh} + w_2 \tilde{n}_{sh} \qquad (5.38a)$$

$$\text{Subject to } m_{i,\min} \leq m_i \leq m_{i,\max} \qquad (5.38b)$$

$$l_{i1,\min} \leq l_{i1} \leq l_{i1,\max} \qquad (5.38c)$$

$$d_{i,\min} \leq d_i \leq d_{i,\min} \qquad (5.38d)$$

$$m_i d_i^2 \leq I_i \qquad\qquad (5.38e)$$

for $i=1, \ldots, n$, where $m_{i,\min}$, $m_{i,\max}$, $l_{i1,\min}$, and $l_{i1,\max}$ are the lower and upper bounds on m_i and l_{i1}, respectively, and $m_i = m_{i1} + m_{i2} + m_{i3}$. The feasibility of the mass center location and the moment of inertia of the ith link can be achieved using constraints, Eqs. (5.38d-e), where $I_i = I_i^c + m_i d_i^2$, which implies that the term $m_i d_i^2$ must be less than or equal to the moment of inertia, I_i.

5.2.6 Counterweighting method

In the case of counterweight balancing, counterweights are attached to the moving links such that the shaking force and shaking moment transmitted to the frame of the mechanism is minimum. Assume the counterweight of mass, m_i^b, with its mass center location, $(\bar{x}_i^b, \bar{y}_i^b)$, is attached to the ith link as shown in Fig. 5.4(a). The equimomental system of the resulting link is shown in Fig. 5.4(b), where it is assumed that the point-masses of the counterweight mass, m_{ij}^b, are placed at the location of the point-masses of original link, m_{ij}^o. Then the counterweight mass, its mass center location and inertia are defined as

$$m_i^b = \sum_{j=1}^{3} m_{ij}^b \qquad\qquad (5.39)$$

$$m_i^b \mathbf{d}_i^b = \sum_{j=1}^{3} m_{ij}^b \mathbf{l}_{ij}^o \qquad\qquad (5.40)$$

$$I_i^b = \sum_{j=1}^{3} m_{ij}^b (l_{ij}^o)^2 \qquad\qquad (5.41)$$

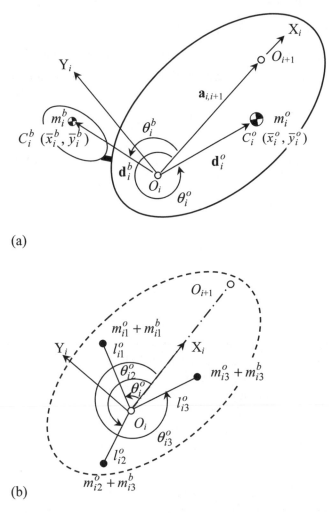

(a)

(b)

Fig. 5.4. Counterweight balancing (a) Counterweight to the ith link (b) Equimomental point-masses of the counterweight mass

Now, for a mechanism having n moving links, the $3n$-vector of the design variables, $\mathbf{x}^b$, is

$$\mathbf{x}^b \equiv [\mathbf{m}_1^{b^T}, ..., \mathbf{m}_n^{b^T}]^T \qquad (5.42)$$

where the 3-vectors, $\mathbf{m}_i^b$, is as follows:

$$\mathbf{m}_i^b \equiv \begin{bmatrix} m_{i1}^b & m_{i2}^b & m_{i3}^b \end{bmatrix}^T, \text{ for } i=1, ..., n.$$

Note that m_{ij}^b is the jth point mass of the counterweight attached to the ith link. The minimum and maximum mass of counterweight, $m_{i,\min}^b$ and $m_{i,\max}^b$, their locations and the moment of inertia depend on an application. However, the counterweight balancing problem is stated to determine the mass, m_i^b, its mass center location, $(\bar{x}_i^b, \bar{y}_i^b)$, and the inertia, I_i^b, such that the combined effect of shaking force and shaking moment is going to be minimum, i.e.,

$$\text{Minimize } z(\mathbf{x}^b) = w_1 \tilde{f}_{sh} + w_2 \tilde{n}_{sh} \tag{5.43a}$$

$$\text{Subject to } m_{i,\min}^b \leq m_i^b \leq m_{i,\max}^b \tag{5.43b}$$

$$d_{i,\min}^b \leq d_i^b \leq d_{i,\max}^b \tag{5.43c}$$

$$m_i^b (d_i^b)^2 \leq I_i^b \tag{5.43d}$$

for $i=1, \ldots, n$, where $d_i^b = \sqrt{\bar{x}_i^{b^2} + \bar{y}_i^{b^2}}$. Similar to the constraints in the mass redistribution method, the mass center location and the moment of inertia of the counterweight attached to the ith link are constraints using inequalities of Eqs. (5.43c-d), respectively.

The optimization methodology using either the mass redistribution and counterweight methods is summarized in the following steps:

1. To harmonize the values of the link parameters, the parameters of the unbalanced mechanism are made dimensionless as explain in Table 5.1.

Table 5.1. Definition of normalized parameters

$a_{ij} =: \lvert \mathbf{a}_{ij} \rvert / a_m$	Normalized distance between joints i and j
$d_i =: \lvert \mathbf{d}_i \rvert / a_m$	Normalized distance of the mass center
$m_i =: m_i / m_m^o$	Normalized mass of the ith link
$I_i =: I_i / (m_m^o a_m^2)$	Normalized moment of inertia of the ith link

The variables, a_m and m_m^o, are defined after Eq. (5.32).

2. Given mass, its mass center location, and the inertia of each link: m_i, $\bar{x}_i$, $\bar{y}_i$, I_i, of the normalized unbalanced mechanism, find the set of

equimomental point-masses for each rigid link. For three point-mass model, using Eq. (5.8), four parameters per link are determined as

$$l_{i1} = \sqrt{I_i / m_i}$$

$$\begin{bmatrix} m_{i1} \\ m_{i2} \\ m_{i3} \end{bmatrix} = \frac{k}{\det(\mathbf{K})} \begin{bmatrix} k\sin(\theta_{i3} - \theta_{i2}) & (\sin\theta_{i2} - \sin\theta_{i3}) & (\cos\theta_{i3} - \cos\theta_{i2}) \\ -k\sin(\theta_{i3} - \theta_{i1}) & (\sin\theta_{i3} - \sin\theta_{i1}) & (\cos\theta_{i1} - \cos\theta_{i3}) \\ -k\sin(\theta_{i1} - \theta_{i2}) & (\sin\theta_{i1} - \sin\theta_{i2}) & (\cos\theta_{i2} - \cos\theta_{i1}) \end{bmatrix} \begin{bmatrix} m_i \\ m_i d_i \cos\theta_i \\ m_i d_i \sin\theta_i \end{bmatrix}$$

where det(K) is defined similar to the one appears after Eq. (4.22), and $\mathbf{d}_i \equiv [\bar{x} \quad \bar{y}]^T$ is defined with respect to the local link-fixed frame.

3. Define design variable for the mechanism having n moving links, as in Eq. (5.37) and Eq. (5.42), for the redistribution and counterweight balancing methods, respectively.

4. Define objective function and the constraints on the link masses and inertias, i.e., Eqs. (5.38a-e) or Eqs. (5.43a-d), where the normalized shaking force and shaking moment are defined according to Eqs. (5.31) and (5.32), respectively. For the normalized mechanism operating at $\omega_{in} = 1$ rad/sec, the shaking force and shaking moment are the normalized shaking force and shaking moment.

5. Solve the optimization problem posed in the above step (4) using any standard optimization technique, say, the optimization toolbox of MATLAB [126]. The optimization process can be started with the parameters of the given unbalanced mechanism as the initial design vector.

6. From the optimized parameters, m_{i1}^*, m_{i2}^*, m_{i3}^*, l_{i1}^*, in redistribution method, the optimized mass, m_i^*, the location of the mass center, ($\bar{x}_i^*, \bar{y}_i^*$), and the inertia of each link, I_i^*, of the balanced mechanism are determined using the equimomental conditions, i.e., Eq. (5.1-3). Similarly, in counterweight method the optimized total mass, m_i^{b*}, the location of the mass center, ($\bar{x}_i^{b*}, \bar{y}_i^{b*}$), and the inertia of counter-weight attached to each link, I_i^{b*}, of the balanced mechanism are de-termined using the equimomental conditions, i.e., Eq. (5.39-41) from optimized point masses, m_{i1}^{b*}, m_{i2}^{b*}, m_{i3}^{b*}.

7. Actual values of link masses or counterweights, their mass center lo-cation, and moments of inertia, are obtained by multiplying the opti-mized values with the corresponding normalizing factors, namely, m_m^o, a_m, and $m_m^o a_m^2$, respectively.

5.3 Hoeken's Four-bar Mechanism

Hoeken's four-bar mechanism is used to generate approximate straight line[133]. It is basically the loop, #0-#1-#2-#3 of Fig. 3.10(b), which is re-drawn in Fig. 5.5 to show the geometry of each link. The link lengths of the mechanism made dimensionless with respect to the parameters of the 1st link. The normalized link parameters of the Hoeken's mechanism used in the carpet scrapping machine [78] are given in Table 5.2. Using the three point-mass model, the equimomental point-masses for each link are obtained using Eqs.(5.1-3), which are given in Table 5.3.

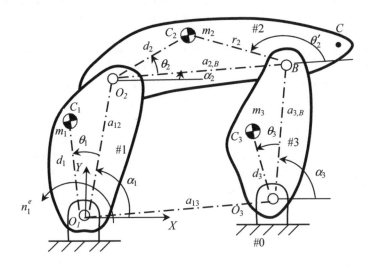

Fig. 5.5. Hoeken's mechanism

Table 5.2. Normalized parameters of the original Hoeken's mechanism

Link i	1	2	3
Link length, a_{ik}	1	3.02	3.02
Link mass, m_i^o	1	7.100	1.900
Center of mass location, d_i^o	0.15	2.86	1.25
Center of mass location, θ_i^o (rad)	0	0	0
Moment of inertia, I_i^o	0.4023	84.4028	8299

Reference parameters: $a_m \equiv O_1O_2 = a_{12} = 0.038$ m; $m_m^o \equiv m_1^o = 0.326$ kg; Length of fixed link, #0: $O_1O_3 = a_{13} = 2.34$ @0°; $\omega_{in} \equiv \omega_1 = 1$ rad/sec.

Table 5.3. Equimomental point-masses of the normalized Hoeken's mechanism

Link i	1	2	3
Point-mass, m_{i1}	0.4910	6.2930	1.5372
Point-mass, m_{i2}	0.2545	0.4035	0.1814
Point-mass, m_{i3}	0.2545	0.4035	0.1814
Distance of point-masses, l_{il}	0.6343	3.4479	1.7517

Other parameters: $\theta_{i1} = 0$; $\theta_{i2} = 2\pi/3$; $\theta_{i3} = 4\pi/3$; $l_{i2} = l_{i3} = l_{i1}$

It is now assumed that no external force and moment act other than the driving torque by the actuator. Then, the shaking force and shaking moment with respect to O_1 that are transmitted to the fixed link, #0, are obtained using Eqs. (5.29-30), respectively, as:

$$\mathbf{f}_{sh} = -(\mathbf{f}_{01} + \mathbf{f}_{03}) \tag{5.44}$$

$$n_{sh} = -(n_{01} + n_{03} - \mathbf{a}_{13}^T \overline{\mathbf{E}} \mathbf{f}_{03} + n_1^e) \tag{5.45}$$

where n_1^e is the driving torque applied at joint 1. Since all joints in the mechanism are revolute, $n_{01} = n_{12} = n_{23} = n_{03} = 0$. In order to know the effect of the force balancing, the optimization problem of the mechanism is first formulated as the force balancing problem in the next subsection.

5.3.1 Balancing of shaking force

The effectiveness of the balancing method proposed in Section 5.2 is compared first with the analytical balancing method proposed by Berkof and Lowen [85]. Berkof and Lowen found the balancing conditions for the total force balance of a general four-bar linkage by redistributing its link masses. The method is based on making the total mass center of the mechanism stationary called as the method of linearly independent vectors. The conditions for the total force balance of the Hoeken's mechanism can be given as follows:

$$\left. \begin{aligned} m_1 d_1 &= m_2 r_2 \frac{a_{12}}{a_{2,B}} \\ \theta_1 &= \theta_2' \end{aligned} \right\} \tag{5.46}$$

$$m_3 d_3 = m_2 d_2 \frac{a_{3,B}}{a_{2,B}} \left.\begin{array}{c} \\ \\ \end{array}\right\}$$
$$\theta_3 = \theta_2 + \pi$$

(5.47)

where point B is the location of the joint between links, #2 and #3, and all the link parameters are shown in Fig.5.5. In Eqs. (5.46-47), there are nine unknowns, d_i, θ_i, and m_i for i=1, 2, 3, and four equations. If the parameters of link 2 are prescribed then angles θ_1 and θ_3 can be found. This determines the radial lines on which the mass centers of the links, #1 and #3, must be placed. Therefore, only the product of mass and distance of the links, #1 and #3, $m_1 d_1$ and $m_3 d_3$, can be found.

To find similar results from the methodology presented in this Chapter, parameters of the link #2 are assumed to be its original ones. The parameters of the remaining two links, #1 and #3, form an 8-vector of design variables, $\mathbf{x}$, as $\mathbf{x} \equiv [m_{11}, m_{12}, m_{13}, l_{11}, m_{31}, m_{32}, m_{33}, l_{31}]^T$. Considering three cases: (a) $\sigma = 1$; (b) $\sigma = 2$; and (c) $\sigma = 5$ with $l_{i1} = l_{i1}^o$ for the analysis, the problem of shaking force balancing is formulated as

$$\text{Minimize } z = \tilde{f}_{sh}$$

(5.48a)

$$\text{Subject to } m_{i1} + m_{i2} + m_{i3} = \sigma m_i^o \text{ , for } i{=}1, 3$$

(5.48b)

where m_i^o and l_{i1}^o are normalized parameters of the original mechanism as given in Table 5.2. In case (a) the masses of links, #1 and #3, are kept same as those in the original mechanism. In cases (b) and (c), the link masses are increased to twice and five times, respectively. Optimization toolbox of MATLAB [126] is used to solve the problem posed in Eq. (5.48). The optimized results for the point-masses, and the geometry of the links obtained from them are given in Table 5.4. The angles, θ_i, and the mass-distance products, $m_i d_i$, for i=1, 3, are shown in the last two columns. The optimized results obtained based on the formulation are in good agreement with the analytical conditions [85] given by Eqs. (5.46-47).

The normalized shaking forces in all the cases are reduced to a negligible value in comparison to that in the original mechanism, as shown in Table 5.5. As expected, the RMS values of the normalized shaking moment, the driving torque, and the bearing reactions at the joints increase with the link masses. For example, in case (c), the shaking moment increases by 38 percent over the corresponding values of the original mechanism. Similarly, Table 5.5 shows the increase in the bearing reactions also.

Table 5.4 Optimized values for the link parameters of Hoeken's mechanism

Cases	Link	Point-masses			Total mass	Mass center location		Mass-distance product
	i	m_{i1}	m_{i2}	m_{i3}	m_i	d_i	θ_i (deg)	$m_i d_i$
(a)	1	-0.0619	0.5309	0.5309	0.9999	0.3761	**180.00**	**0.3760**
	3	-7.0949	4.4975	4.4975	1.9001	10.6870	**180.00**	**20.3064**
(b)	1	0.2715	0.8643	0.8643	2.0001	0.1880	**180.00**	**0.3760**
	3	-6.4616	1308	1308	3.8000	3438	**180.00**	**20.3064**
(c)	1	1.2715	1.8643	1.8643	0001	0.0752	**180.00**	**0.3760**
	3	-4.5616	7.0308	7.0308	9.5000	2.1375	**180.00**	**20.3064**

Values using Berkof and Lowen[85] conditions: $m_1 d_1 = 0.3762$; $m_3 d_3 = 20.3060$ and $\theta_1 = 180°$; $\theta_3 = 180°$.

Table 5.5. RMS values of the dynamic quantities for the Hoeken's mechanism

Mechanism	Bearing reaction forces				Input torque	Sh. force	Sh. moment
	$\tilde{f}_{01}$	$\tilde{f}_{12}$	$\tilde{f}_{23}$	$\tilde{f}_{03}$	$\tilde{\tau}$	$\tilde{f}_{sh}$	$\tilde{n}_{sh}$
Original	24.778	24.693	22.932	23.039	1317	11.536	36.779
Case (a)	24.481	24.693	22.932	24.481	1317	0.000	40.238
					(00)	(-100)	(09)
Case (b)	26.131	26.342	24.335	26.131	16.356	0.000	42.899
					(07)	(-100)	(17)
Case (c)	31.100	31.310	28.724	31.100	19.486	0.000	50.899
					(27)	(-100)	(38)

The values in the parenthesis denote the percentage increase or decrease over the corresponding values for the original mechanism.

5.3.2 Optimization of shaking force and shaking moment

The effect of force balancing on the bearing reactions, driving torque, and the shaking moment in a family of four-bar linkages is investigated by Lowen et al. [87]. They had shown that all the mentioned dynamic quantities increase up to 50 percent in most of the cases. This fact is also clear from the results of the previous subsection. The conclusions of Lowen et al. [87] and other researchers suggest that only the force balancing is not effective, and the balancing of mechanisms requires trade-off amongst various dynamic quantities. In this section, the optimality criteria given in Eq. (5.34) is used which combine both the shaking force and shaking moment. The suitable constraints on the design variables are also considered.

In this study, two sets of constraints on the design variables are considered in the mass redistribution method, namely, (a) total mass of each link, its mass center location and moment of inertia, and the distances of the point-masses from the link's origin are constrained. Here negative values of the point-masses are allowed; (b) In addition to the constraints of (a), negative values of the point-masses are not allowed. For both the problems, $\theta_{i1} = 0$, $\theta_{i2} = 2\pi/3$, $\theta_{i3} = 4\pi/3$ and $l_{i2} = l_{i3} = l_{i1}$ are taken as pre-assigned parameters, and m_{i1}, m_{i2}, m_{i3}, and l_{i1}, for i=1, 2, 3 are the design variables. Then, the design vector, $\mathbf{x}$, consists of a total of 12 design variables, m_{ij} and l_{i1}, for i, j=1, 2, 3, which is defined as

$$\mathbf{x} \equiv [m_{11}, m_{12}, m_{13}, l_{11}, m_{21}, m_{22}, m_{23}, l_{21}, m_{31}, m_{32}, m_{33}, l_{31}]^{T}.$$

In the counterweight method, it is assumed that the mechanism is balanced by two counterweighs attached to the links, #1 and #3. Hence, the design vector, $\mathbf{x}^{b}$, consists of 6 design variables, m_{ij}^{b} , for i=1, 3, and j=1, 2, 3, i.e.,

$$\mathbf{x}^{b} \equiv [m_{11}^{b}, m_{12}^{b}, m_{13}^{b}, m_{31}^{b}, m_{32}^{b}, m_{33}^{b}]^{T}.$$

Note here that the locations of the point masses of counterweight of each link are fixed in the body-fixed frame as they are assumed to be located at the positions of the point-masses of the original link. The balancing problem is then posed as

Minimize $z = w_1 \tilde{f}_{sh} + w_2 \tilde{n}_{sh}$ (5.49)

(*i*) For mass redistribution method

(a) Subject to $m_i^o \leq m_i \leq 5m_i^o$ (5.50a)

$0.5l_i^o \leq l_{i1} \leq 1.5l_i^o$ (5.50b)

$d_i \leq a_i$ (5.50c)

$m_i d_i^2 \leq I_i$ for i=1, 2, 3 (5.50d)

(b) Subject to $m_i^o \leq m_i \leq 5m_i^o$ (5.51a)

$$0.5l_i^o \leq l_{i1} \leq 1.5l_i^o$$ (5.51b)

$$d_i \leq a_i$$ (5.51c)

$$m_i d_i^2 \leq I_i$$ (5.51d)

$$m_{i1}, m_{i2}, m_{i3} \geq 0 \text{ for } i=1, 2, 3$$ (5.51e)

(*ii*) For counterweight balancing method

(c) Subject to $0 \leq m_i^b \leq 5m_i^o$ (5.52a)

$$d_i^b \leq a_i$$ (5.52b)

$$m_i^b(d_i^b)^2 \leq I_i^b \text{ for } i=1, 3$$ (5.52c)

Here a_i is length of the ith link. Choosing three sets of weighting factors (w_1, w_2): (1, 0), (0.5, 0.5), and (0, 1), a total of nine cases are investigated. Table 5.6 shows the optimal design vectors, $\mathbf{x}^*$ and $\mathbf{x}^{b*}$, obtained for each case in the mass redistribute and counterweight methods. The geometry and inertial properties of the mechanism for each case are obtained back from the design vector using Eqs. (5.1-3), and given in Table 5.7. Table 5.8 shows the comparison between the RMS values of the normalized dynamic quantities of the balanced mechanism occurred during the complete motion cycle with those of the normalized original mechanism. The following conclusions are accrued from the comparison of the results given in Tables 5.6-5.8:

- The RMS values of the shaking force shaking moment and the driving torque occurred during the motion cycle are minimum when equal weight is given to both the quantities, as shown by the bold numbers in Table 5.8.
- The mass redistribution method is more effective in comparison to the counterweight method. The reductions in shaking force and shaking moment are 89% and 87%, respectively, in case a(2), whereas only10%, and 7% in case c(2).
- Comparison of two cases a(2) and b(2) shows that the results obtained with condition that negative point-masses are allowed are better. It is not

necessary that the point-masses should be positives. The condition for a set of positive and negative point-masses is that they should represent a link where the total mass and moment of inertia about the axes through the center of mass must be positive. Non-negativity of the total mass of each link is achieved by constraining it as, $m_i^o \leq m_i \leq 5m_i^o$, whereas the non-zero moment of inertia is achieved by assuming, $l_{i2} = l_{i3} = l_{i1}$ and restricting l_{i1} as $0.5l_i^o \leq l_{i1} \leq 1.5l_i^o$.

- The mechanism geometry obtained in case a(1) for shaking force balancing also satisfy the analytical conditions, Eqs. (5.46-47) given by Berkof and Lowen [85]. Note that, to use these conditions, one needs to assume the geometry of the coupler link, whereas no such assumption is required in the optimization method given here. It optimizes distribution of the masses of all the links.

Table 5.6. Design vectors for the balanced Hoeken's mechanism

Case: (w_1, w_2)	Design vector
(*i*) Mass redistribution method:	
$\mathbf{x} \equiv [m_{11}, m_{12}, m_{13}, l_{11}, \; m_{21}, m_{22}, m_{23}, l_{21}, m_{31}, m_{32}, m_{33}, l_{31}]^T$	
a(1): (1.0, 0.0)	[-0.1530 2.2957 2.8324 1.6897 4319 0.0159 1.6662 1.6595 -1.0874 3.6672 2.2184 1.8904]^T
a(2): (0.5, 0.5)	[-1.4604 2.2304 4.2300 0.9514 4.8672 -1.0361 3.2689 1.7240 -0.2608 9.4926 0.2682 0.9165]^T
a(3): (0.0, 1.0)	[0.4910 0.2545 0.2545 0.6343 4.9288 0.4033 1.7679 1.7240 2.6474 8.8926 -2.0401 1.2391]^T
b(1): (1.0 ,0.0)	[0 2.3350 2.6650 0.9514 9063 0.3106 0.8831 2.4240 0 4.9305 4.5695 2.6276]^T
b(2): (0.5 ,0.5)	[0 1.5989 3.4011 0.9514 4.6933 0 2.4067 1.7240 0 9.5000 0 0.8759]^T
b(3): (0.0, 1.0)	[0.4910 0.2545 0.2545 0.6343 4.9671 0 2.1329 1.7240 0 9.5000 0 0.9369]^T
(*ii*) Counterweight method:	
$\mathbf{x}^b \equiv [m_{11}^b, m_{12}^b, m_{13}^b, \; m_{31}^b, m_{32}^b, m_{33}^b]^T$	
c(1): (1.0 ,0.0)	[0.4910 0.2545 0.2545 3.6642 8.3858 -2.5500]^T
c(2): (0.5 ,0.5)	[-1.6533 3.5852 3.0681 0.9971 8.5245 -0.8927]^T
c(3): (0.0 ,1.0)	[-0.6433 2.6499 0.8661 -3.1667 6.3337 6.3329]^T

Table 5.7. Link parameters of the balanced Hoeken's mechanism

Case	Link, i	Total link mass m_i^*	Mass center location d_i^*	θ_i^* (deg)	Moment of inertia I_i^*
(i) Mass redistribution method					
Case: a(1)	1	4.975	0.936	189.71	14.204
	2	7.114	1.122	342.71	19.592
	3	4.798	1.663	162.71	17.147
		16.887[+]			
Case: a(2)	1	000	0.951	200.26	4.526
	2	7.100	1.284	3117	21.102
	3	9.500	0.917	122.76	7.980
		21.600			
Case: a(3)	1	1.000	0.150	0.00	0.402
	2	7.100	0.976	342.91	21.102
	3	9.500	1.239	94.70	14.586
		17.600			
Case: b(1)	1	000	0.479	186.52	4.526
	2	7.100	1.821	354.67	41.718
	3	9.500	1.317	176.23	6591
		21.600			
Case: b(2)	1	000	0.561	211.98	4.526
	2	7.100	0.987	329.15	21.102
	3	9.500	0.876	120.00	7.288
		21.600			
Case: b(3)	1	1.000	0.150	0.00	0.402
	2	7.100	1.048	334.66	21.102
	3	9.500	0.937	120.00	8.339
		17.600			
(ii) Counterweight method :		m_i^{b*}	d_i^{b*}	θ_i^{b*}	I_i^{b*}
Case: c(1)	1	2.873	0.630	147.25	1.156
	3	9.500	1.752	180.00	29.150
		12.373[++]			
Case: c(2)	1	000	0.634	174.86	2.012
	3	8.629	1.752	109.07	26.477
		13.629			
Case: c(3)	1	1.000	0.150	0.00	0.402
	3	9.500	1.752	849	29.150
		10.500			

[+]Σm_i^* ; [++]Σm_i^{b*} ; and total mass of the original mechanism, $\Sigma m_i^o = 10.000$

Table 5.8. RMS values for the Hoeken's mechanism

Case	Bearing reactions				Input torque	Sh. force	Sh. moment
	$\tilde{f}_{01}$	$\tilde{f}_{12}$	$\tilde{f}_{23}$	$\tilde{f}_{03}$	$\tilde{\tau}$	$\tilde{f}_{sh}$	$\tilde{n}_{sh}$
Original mechanism	24.778	24.693	22.932	23.038	1317	11.536	36.778
(*i*) Mass redistribution method							
Case: a(1)	9.558	12.473	8.739	9.558	924 (-61)*	0.000 (-100)	14.019 (-62)
Case: a(2)	**457**	**8.456**	**693**	**4.708**	**2.701 (-82)**	**1.322 (-89)**	**4.647 (-87)**
Case: a(3)	12.385	12.267	8.504	4.921	535 (-64)	9.165 (-21)	2.703 (-93)
Case: b(1)	30.174	31.521	27.774	29.977	18.804 (23)	0.521 (-95)	47.938 (30)
Case: b(2)	**7.779**	**9.575**	**6.187**	**4.575**	**3.441 (-78)**	**4.049 (-65)**	**4.240 (-88)**
Case: b(3)	10.111	9.987	6.532	4.856	3.954 (-74)	6.691 (-42)	4.176 (-89)
(*ii*) Counterweight method							
Case: c(1)	32.238	32.970	30.232	31.691	20.532 (34)	2.677 (-78)	52.289 (42)
Case: c(2)	**30.628**	**32.208**	**29.539**	**23.803**	**20.052 (31)**	**10.437 (-10)**	**34.096 (-7)**
Case: c(3)	33.139	32.970	30.232	23.299	20.532 (34)	14.919 (29)	32.225 (-12)

*The values in the parenthesis denote the round-off percentage increase or decrease over the corresponding values for the original mechanism.

Figures 5.6-5.8 show a comparison of the dynamic quantities with those of the original mechanism. The figures clearly demonstrate that cases, a(2) and b(2) of the mass redistribution method, and c(2) of counterweight method, are effective. Figures 5.9 shows the location of link mass centers of the balanced mechanism obtained in case a(2).

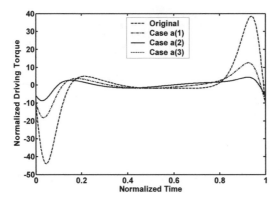

(*i*) Driving torque

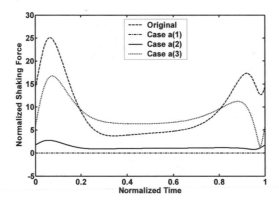

(*ii*) Shaking force

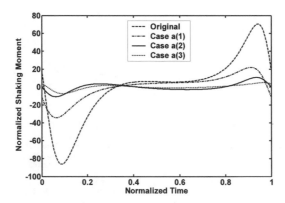

(*iii*) Shaking moment

Fig. 5.6. Dynamic performances of the Hoeken's mechanism, case (a)

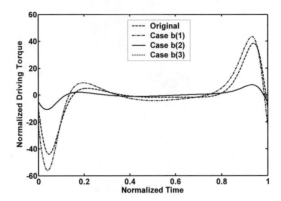

(*i*) Driving torque

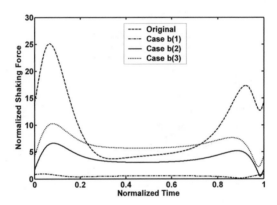

(*ii*) Shaking force

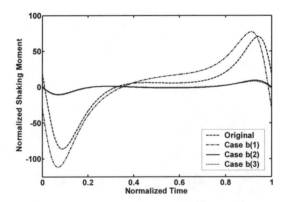

(*iii*) Shaking moment

Fig. 5.7. Dynamic performances of the Hoeken's mechanism, case (b)

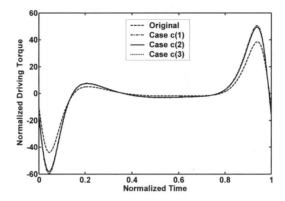

(*i*) Driving torque

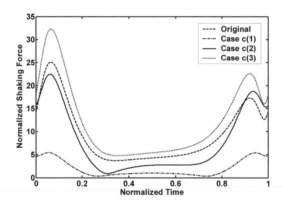

(*ii*) Shaking force

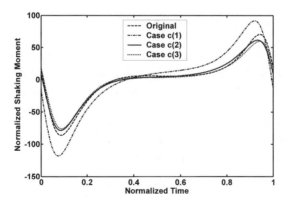

(*iii*) Shaking moment

Fig. 5.8. Dynamic performances of the Hoeken's mechanism, case (c)

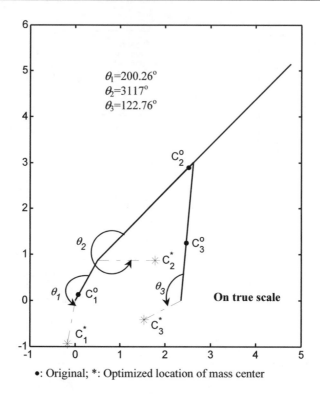

$\bullet$: Original; $*$: Optimized location of mass center

Fig. 5.9. Locations of link mass centers of the balanced normalized Hoeken's mechanism, case a(2)

5.4 Carpet Scrapping Mechanism

The carpet scrapping mechanism shown in Fig. 3.10(b) is used here to show the applicability of the optimization method to multiloop mechanisms. The mechanism parameters are normalized with respect to the parameters of the first link and its angular velocity. The geometry, mass, and inertia of the normalized original mechanism are given in Table 5.9. The shaking force and shaking moment transmitted to the fixed link #0, using Eqs. (5.29-30), respectively, are given by

$$\mathbf{f}_{sh} = -(\mathbf{f}_{01} + \mathbf{f}_{03} + \mathbf{f}_{04})$$

(5.53)

$$n_{sh} = -(n_{01} + n_{03} + n_{04} - \mathbf{a}_{13}^T\overline{\mathbf{E}}\mathbf{f}_{03} - \mathbf{a}_{14}^T\overline{\mathbf{E}}\mathbf{f}_{04} + n_1^e) \tag{5.54}$$

where n_1^e is driving torque applied at joint between links #0 and #1, as in Fig. 3.10(b). Dynamic analysis for the computation of the reactions is shown in Section 3.4, while the dynamic equations in terms of the design variables are shown in Subsection 5.2.1. The parameters, θ_{i1}, θ_{i2}, and θ_{i3} are equal to 0, $2\pi/3$, $4\pi/3$, respectively, and $l_{i2} = l_{i3} = l_{i1}$. Hence, the 28-dimesional design vector, $\mathbf{x}$, takes the following form:

$$\mathbf{x} \equiv [m_{11}, m_{12}, m_{13}, l_{11}, \cdots, m_{71}, m_{72}, m_{73}, l_{71}]^T.$$

Table 5.9. Normalized parameters of the original carpet scrapping mechanism

Link i	1	2	3	4	5	6	7
Link length, a_{ik}	1	3.02	3.02	8.78	6.27	6.27	6.27
Link mass, m_i^o	1	7.100	1.900	200	4.300	14.200	4.300
Mass center location, d_i^o	0.15	2.86	1.25	4.50	3.14	10.00	3.14
Mass center location, θ_i^o (deg)	0	0	0	0	0	0	0
Moment of inertia, I_i^o	0.4023	84.4028	8299	143.6004	61.5979	2007.2	61.5979

Reference link: $a_m \equiv a_{12} = 0.038$ m; $m_m^o \equiv m_1^o = 0.3260$ kg; $O_1O_3 = 2.34$ @180°; $O_1O_4 = 10.79$ @84.7°; and $\omega_1 = 1$ rad/sec.

Constraining the link masses, m_i, between m_i^o and $5m_i^o$, and the variable, l_{i1}, between $0.5l_{i1}^o$ and $1.5l_{i1}^o$, the optimization problem for the scrapping mechanism is posed as

Minimize $z = w_1\tilde{f}_{sh} + w_2\tilde{n}_{sh}$ $\quad$ (5.55a)

Subject to $m_i^o \le m_i \le 5m_i^o$ $\quad$ (5.55b)

$0.5l_i^o \le l_{i1} \le 1.5l_i^o$ $\quad$ (5.55c)

$d_i \le a_i^o$ $\quad$ (5.55d)

$$m_i d_i^2 \leq I_i \text{ for } i=1, \dots, 7 \tag{5.55e}$$

The feasibility of the mass center location and the moment of inertia of each link is achieved using constraints, Eqs. (5.55c-e). Based on the conclusions given in Section 5.3.2, the point masses are allowed to take negative values. The whole algorithm is coded in the MATLAB to determine the time dependent behavior of the various relevant quantities, including the bearing forces and driving torque.

The results obtained by applying different weighting factors, w_1 and w_2, are given in Table 5.10. These weighting factors are nothing but the importance given to different competing quantities. The values depend on the application and can be generally considered equal to each other [115]. Three cases are investigated using three sets of the weighting factors, i.e., $(w_1, w_2)=$ (1.0, 0.0); (0.5, 0.5); and (0.0, 1.0). The optimized mass, its mass center location, and the moment of inertia of each link is then calculated from optimized point-masses using Eqs. (5.1-3), which are shown in Table 5.11 for case (2). The RMS values of the dynamic quantities of the balanced mechanism are compared in Table 5.12 with those corresponding to the original mechanism. The results show a significant improvement in the dynamic performances. For example, in case (2), a reduction of 15, 58, and 67 percent is observed in the RMS values of the driving torque, shaking force and shaking moment, respectively. Figure 5.10 shows the comparison between the dynamic performances of the mechanism. Locations of link mass centers of the balanced normalized carpet scrapping mechanism for case (2) are shown in Fig. 5.11.

Table 5.10. Design vectors for the balanced scrapping mechanism

Case: (w_1, w_2)	Design vector, x^*
Case (1): (1.0,0.0)	[-1.6586 3.1290 3.5296 0.9514 5467 1.5138 0.0396 3.5518 1.1957 9.3646 -1.0603 2.6276 -7.2980 15490 14.3853 7.8663 -0.2885 0.8307 3.7578 3.8787 0.3706 8.3873 9176 7.2007 3.0513 1.0887 0.1602 4.4101]^T
Case (2): (0.5,0.5)	[0.0827 -0.0813 4.9987 0.9514 -3.3751 5043 8.7254 2.5783 -3.1224 6.9588 6636 2.6276 -8.4745 10080 19.4664 2.6275 1.5700 -1.1148 3.8448 6773 12.3363 -2.5706 4.4344 6.8925 4.2982 -0.0881 0.0899 6773]^T
Case (3): (0.0,1.0)	[0.4910 0.2545 0.2545 0.6343 30.1479 -6.1652 11.5173 3.4087 -3.0471 2126 7.3344 2.6276 -8.1689 13.5128 20.6561 2.6275 1.6690 -1.2780 4557 6773 12.4356 0.3290 1.4354 7.6799 4.2987 0.0760 -0.0747 6773]^T

Table 5.11. Link parameters of the balanced crapping mechanism

Case	Link, i	Total link mass m_i^*	Mass center location d_i^*	θ_i^* (deg)	Moment of inertia I_i^*
Case (2)	1	000	0.951	-118.37	4.526
	2	10.855	2.578	-1611	72.157
	3	9.500	2.628	173.22	6591
	4	26.000	2.628	-171.46	179.497
	5	4.300	677	-87.27	138.596
	6	14.200	6.270	-28.01	674.598
	7	4.300	677	-2.05	138.596
		74.155$^+$			

$^+\Sigma m_i^*$; and total mass of the original mechanism, $\Sigma m_i^o = 38.000$

Table 5.12. RMS values for the scrapping mechanism

Case	Input torque $\tilde{\tau}$	Sh. force $\tilde{f}_{sh}$	Sh. moment $\tilde{n}_{sh}$
Original	1192.271	182.399	413.496
Case: (1) $w_1=1.0;w_2=0.0$	2657.200 (123)	4.400 (-98)	3392.200 (720)
Case: (2) $w_1=0.5;w_2=0.5$	**1013.200** (-15)	**77.000** (-58)	**138.400** (-67)
Case: (3) $w_1=0.0;w_2=1.0$	1178.200 (-01)	148.800 (-18)	117.900 (-71)

The values in parenthesis denote round-off percentage increase or decrease over the corresponding values for the original mechanism

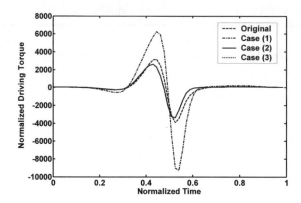

(*i*) Driving torque

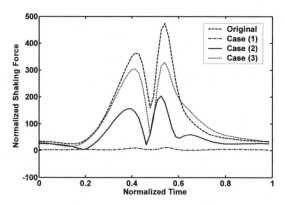

(*ii*) Shaking force

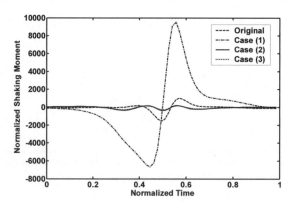

(*iii*) Shaking moment

Fig. 5.10. Dynamic performances of the carpet scrapping mechanism

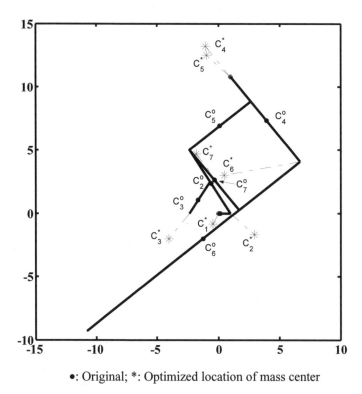

•: Original; *: Optimized location of mass center

Fig. 5.11. Locations of link mass centers of the balanced normalized carpet scrapping mechanism, case (2)

5.5 Summary

In this chapter, balancing problem of planar mechanisms is formulated as an optimization problem. The main focus of the chapter is to reduce the shaking force and shaking moment. The design variables and the constraints on them are identified by introducing the equimomental system of point-masses. The three point-mass model is used for the planar mechanisms. Using the equimomental point-masses, equations of motion are reformulated to determine the shaking force, shaking moment, and other dynamic quantities. Two practical mechanisms, one is Hoeken's mechanism and other one is the multiloop carpet scrapping mechanism, are balanced using the methodology given in this Chapter.

6 Balancing of Spatial Mechanisms

The methodology proposed in Chapter 5 to balance planar mechanisms is extended in this chapter for the spatial mechanisms. Spatial mechanisms need more exhaustive treatment compared to its planar counterparts. Both the kinematics and dynamics of the spatial mechanisms are more complex. Regarding the balancing of shaking force and shaking moment in spatial high speed mechanisms/machines, Kaufman and Sandor [103] presented a complete force balancing of spatial mechanisms like RSSR (Revolute-Spherical-Spherical-Revolute) and RSSP (Revolute-Spherical-Spherical-Prismatic). Their approaches are based on the generalization of the planar mechanism balancing theory developed by Berkof and Lowen [85] using a technique of linearly independent vectors. Using the real vectors and the concept of retaining the stationary center of total mass, Bagci [104] has obtained the design equations for force balancing of various mechanisms, whereas Ning-Xin Chen [105, 106] extended the concept of linearly independent vectors to a single loop spatial n-bar linkages with some restricted kinematic pairs for the derivation of the shaking force balance conditions. In addition, shaking moment balancing can be achieved by adding the dyads [107, 108] and rotating mass balancers [109]. However, these balancing methods use counterweights and/or dyads that are restricted to few specific mechanisms only. Also, the techniques like genetic algorithm, etc., were also applied to the optimum balancing of shaking force and shaking moment for the spatial RSSR mechanism [120].

The dynamic behavior of a mechanism depends on its total mass, mass center location, and the inertia tensor. For spatial mechanisms, a system of point-masses is equimomental to the mechanism, if it satisfies the conditions of the same mass, the same mass center location, and the same inertia tensor. Hokey and Sherwood [113] proposed a ten point-mass model to represent the mass and inertial properties of each link. However, they did not apply them to balance any spatial mechanism. Later, Gill and Freudenstein [118], and Rahman [119] proposed a four point-mass model to balance spatial four-bar mechanisms. But their results do not show any significant improvement in reducing the shaking force and shaking moment. This is due to the less number of point-masses and only positive point-masses allowed. A seven point-mass model is proposed here to represent the mass and inertia of each link. Accordingly, the dynamic formulation given in Chapter 3 is reformulated in terms of the point-masses. The methodology is illustrated using the spatial RSSR mechanism.

6.1 Balancing Problem Formulation

As explained in chapter 5, the balancing problem of a mechanism is to find the mass, the location of the mass center, and the inertia of each link so that the shaking force and shaking moment transmitted to the frame of the mechanism are minimum. Hence, it is treated as an optimization problem, which is formulated in this section. To do so, first the Newton-Euler (NE) equations of motion are derived in the parameters of point-masses, which are identified as the *design variables* (DVs). The shaking force and shaking moment are then computed using the constraint force formulation proposed in Chapter 3.

6.1.1 Dynamic equations of motion

Referring to the ith rigid link of a spatial mechanism, Fig. 6.1, points O_i and O_{i+1} are the origins of the coordinate frames, $O_iX_iY_iZ_i$ (F_i) and $O_{i+1}X_{i+1}Y_{i+1}Z_{i+1}$ (F_{i+1}), respectively, as described in Appendix A. Note that the frames, F_i and F_{i+1}, are fixed to the $(i\text{-}1)$st and ith link, respectively. Now, the Newton-Euler (NE) equations of motion [60], as derived in Section 2.3, are given by

$$\mathbf{M}_i\dot{\mathbf{t}}_i + \mathbf{W}_i\mathbf{M}_i\mathbf{E}_i\mathbf{t}_i = \mathbf{w}_i \qquad (6.1)$$

where all the vectors and the matrices are defined after Eq. (2.31).

In order to represent the mass matrix, $\mathbf{M}_i$ of Eq. (6.1), in terms of the parameters of the point-masses, the ith rigid link is modeled as seven equimomental point-mass system, as presented in Section 4.2. Referring to Fig. 6.1, the point-masses are placed at the corners of a rectangular parallelepiped whose center is point O_{i+1}, and the sides are $2h_{ix}$, $2h_{iy}$, $2h_{iz}$ along the axes, X_{i+1}, Y_{i+1} and Z_{i+1}, respectively. The point-masses, m_{i1}, ..., m_{i7}, are fixed to the local frame, i.e., $O_{i+1}X_{i+1}Y_{i+1}Z_{i+1}$, which is attached to the ith link. The 3-vectors, $\mathbf{d}_{ij}$ and $\mathbf{r}_{ij}$, are the positions of the point-mass, m_{ij}, from the origins O_i and O_{i+1}, respectively. Subscripts i and j denote ith link and its jth point-mass, respectively. The components of the vectors, $\mathbf{r}_{ij}$, in the body fixed frame, $O_{i+1}X_{i+1}Y_{i+1}Z_{i+1}$, are given in Table 6.1.

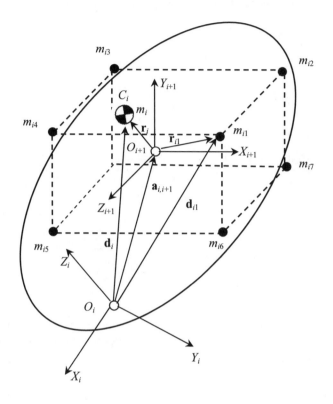

Fig. 6.1. Seven point-masses model

Table 6.1. Components of vectors $\mathbf{r}_{ij}$ in the $(i+1)$st frame

	$[\mathbf{r}_{i1}]_{i+1}$	$[\mathbf{r}_{i2}]_{i+1}$	$[\mathbf{r}_{i3}]_{i+1}$	$[\mathbf{r}_{i4}]_{i+1}$	$[\mathbf{r}_{i5}]_{i+1}$	$[\mathbf{r}_{i6}]_{i+1}$	$[\mathbf{r}_{i7}]_{i+1}$
Along X_{i+1}	h_{ix}	h_{ix}	$-h_{ix}$	$-h_{ix}$	$-h_{ix}$	h_{ix}	h_{ix}
Along Y_{i+1}	h_{iy}	h_{iy}	h_{iy}	h_{iy}	$-h_{iy}$	$-h_{iy}$	$-h_{iy}$
Along Z_{i+1}	h_{iz}	$-h_{iz}$	$-h_{iz}$	h_{iz}	h_{iz}	h_{iz}	$-h_{iz}$

Using the equimomental conditions, Eqs. (4.30-32), vector $\mathbf{d}_i$ denoting the mass center in terms of $\mathbf{d}_{ij}$'s is obtained as

$$\mathbf{d}_i = \frac{1}{m_i} \sum_{j=1}^{7} m_{ij} \mathbf{d}_{ij} \qquad (6.2)$$

where, $\mathbf{d}_{ij} \equiv \mathbf{a}_{i,i+1} + \mathbf{r}_{ij}$, in which $\mathbf{a}_{i,i+1}$ is the link length vector defined from O_i to O_{i+1}, and $\mathbf{r}_{ij}$ is given in Table 6.1. Furthermore, the total mass of the ith link is

$$m_i = \sum_{j=1}^{7} m_{ij} \tag{6.3}$$

Denoting $\mathbf{d}_{ij} \equiv [d_{ijx}, d_{ijy}, d_{ijz}]^T$, the 3×3 skew-symmetric matrix, $\tilde{\mathbf{d}}_i$, associated with the vector, $\mathbf{d}_i$, is represented as

$$\tilde{\mathbf{d}}_i = \frac{1}{m_i} \begin{bmatrix} 0 & -\sum_{j=1}^{7} m_{ij} d_{ijz} & \sum_{j=1}^{7} m_{ij} d_{ijy} \\ \sum_{j=1}^{7} m_{ij} d_{ijz} & 0 & -\sum_{j=1}^{7} m_{ij} d_{ijx} \\ -\sum_{j=1}^{7} m_{ij} d_{ijy} & \sum_{j=1}^{7} m_{ij} d_{ijx} & 0 \end{bmatrix} \tag{6.4}$$

Using the conditions of equality for each component of the inertia tensor, Eqs. (4.33-38), the inertia tensor, $\mathbf{I}_i$, about the origin, O_i, in terms of the point mass parameters has the following representation:

$$\mathbf{I}_i = \begin{bmatrix} \sum_{j=1}^{7} m_{ij}(d_{ijy}^2 + d_{ijz}^2) & -\sum_{j=1}^{7} m_{ij} d_{ijx} d_{ijy} & -\sum_{j=1}^{7} m_{ij} d_{ijx} d_{ijz} \\ & \sum_{j=1}^{7} m_{ij}(d_{ijz}^2 + d_{ijx}^2) & -\sum_{j=1}^{7} m_{ij} d_{ijy} d_{ijz} \\ Sym & & \sum_{j=1}^{7} m_{ij}(d_{ijx}^2 + d_{ijy}^2) \end{bmatrix} \tag{6.5}$$

Equations (6.3-5) define the mass matrix, $\mathbf{M}_i$, of the ith link in terms of the parameters of the equimomental seven point masses. The $6n$ unconstrained scalar equations, Eq. (6.1), for the whole system having n moving bodies, i.e., $i=1, \ldots, n$, are written next in compact form as

$$\mathbf{M}\dot{\mathbf{t}} + \mathbf{WME}\mathbf{t} = \mathbf{w} \tag{6.6}$$

where all the vectors and the matrices are defined after Eq. (2.34). The expressions in the left hand side of Eq. (6.6) denote the effective inertial wrenches, and those on the right hand side represent the external wrenches and those due to the constraints at the joints. In the next step, the shaking force and shaking moment are calculated using the constrained equations of motions that have been formulated in Chapter 3.

6.1.2 Shaking force and shaking moment

Using the definitions given in Section 5.2.3, the shaking force and shaking moment in a mechanism having n moving links are given by

$$\mathbf{f}_{sh} = -\sum_{i=1}^{n} \mathbf{f}_i^* \tag{6.7}$$

$$\mathbf{n}_{sh} = -\sum_{i=1}^{n} (\mathbf{n}_i^* + \tilde{\mathbf{a}}_{1,i}\mathbf{f}_i^*) \tag{6.8}$$

where $\mathbf{f}_i^*$ and $\mathbf{n}_i^*$ are the 3-vectors of inertia force and inertia moment of the ith body acting at and about origin O_i, respectively. The 3×3 matrix, $\tilde{\mathbf{a}}_{1,i}$, is the skew-symmetric matrix corresponding to the 3-vector, $\mathbf{a}_{1,i}$, from O_1 to O_i. The point O_1 is the origin of the frame, $X_1Y_1Z_1$, attached to the fixed link about which the shaking moment is defined. It is evident from Eq. (6.8) that the shaking moment is a pure torque for the fully force balanced mechanism. Referring to Fig. 6.2, the equilibrium of forces and moments are expressed as

$$\mathbf{f}_i^* = \mathbf{f}_i^e + \mathbf{f}_{i-1,i} - \mathbf{f}_{i,i+1} \tag{6.9}$$

$$\mathbf{n}_i^* = \mathbf{n}_i^e + \mathbf{n}_{i-1,i} - \mathbf{n}_{i,i+1} + \tilde{\mathbf{a}}_{i,i+1}\mathbf{f}_{i,i+1} \tag{6.10}$$

where the 3-vectors, $\mathbf{f}_{i-1,i}$, $\mathbf{n}_{i-1,i}$, and $\mathbf{f}_{i,i+1}$, $\mathbf{n}_{i,i+1}$, are the constraint forces and moments at the origins, O_i and O_{i+1}, respectively, whereas the 3-vetors, $\mathbf{f}_i^e$ and $\mathbf{n}_i^e$, are the external force and moment acting on the ith body at and about O_i, respectively. It is assumed that the links, $j=1, \ldots, n_f$, are connected to the fixed link, #0. Then, substituting, Eqs. (6.9-10) into Eqs. (6.7-8), the shaking force and shaking moment, respectively, are obtained as

$$\mathbf{f}_{sh} = -\sum_{j=1}^{n_f} \mathbf{f}_{0,j} - \sum_{i=1}^{n} \mathbf{f}_i^e \tag{6.11}$$

$$\mathbf{n}_{sh} = -\sum_{j=1}^{n_f}(\mathbf{n}_{0,j} + \tilde{\mathbf{a}}_{1,j}\mathbf{f}_{0,j}) - \sum_{i=1}^{n}(\mathbf{n}_i^E + \tilde{\mathbf{a}}_{1,i}\mathbf{f}_i^e) \tag{6.12}$$

where $\mathbf{f}_{0,j}$ represents the reaction force of the fixed link, #0, on the jth link connected to it. For all the other links that are not connected to the fixed one, the term, $\mathbf{f}_{0,j}$, vanishes.

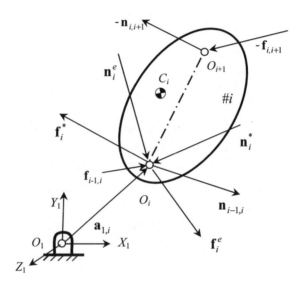

Fig. 6.2. Free body diagram of the ith link

6.1.3 Optimization problem

In this section, two optimization techniques are developed to reduce shaking force and shaking moment: (*i*) redistribution of the mass of moving links, and (*ii*) counterweighting the moving links.

Mass redistribution method

The shaking force and shaking moment are the resultant of the inertia forces and moments of moving links. When the dimensions and the input speed of a mechanism are given, the inertia forces depend only upon the mass distribution of the moving links. Hence, mass redistribution is the obvious choice to balance mechanisms. First, the inertial properties of the links are represented using the seven point-mass model. The seven point masses of each link, m_{i1}, ..., m_{i7}, which are located at the corners of a rectangular parallelepiped of sides, $2h_{ix}$, $2h_{ix}$, and $2h_{iz}$ as explained in Section 4.2, are taken as the design variables. The values of h_{ix}, h_{ix}, and h_{iz} are calculated from the moments of inertia of each original link using Eqs. (4.42-45) and remain constant during the optimization process. For a mechanism having n moving links, the $7n$-vector of the design variables, $\mathbf{x}$, is then defined as

$$\mathbf{x} \equiv [\mathbf{m}_1^T, ..., \mathbf{m}_n^T]^T \qquad (6.13)$$

where $\mathbf{m}_i$ is 7-vectors of point-masses and is defined as

$$\mathbf{m}_i \equiv [m_{i1} \quad m_{i2} \quad m_{i3} \quad m_{i4} \quad m_{i5} \quad m_{i6} \quad m_{i7}]^T$$

Taking the root mean square (RMS) values of the normalized shaking force, $\tilde{f}_{sh}$, and the normalized shaking moment, $\tilde{n}_{sh}$, an optimality criterion is proposed similar to Eq. (5.34) for planar motion as

$$z = w_1 \tilde{f}_{sh} + w_2 \tilde{n}_{sh} \qquad (6.14)$$

where w_1 and w_2 are the weighting factors whose values may vary depending on an application and $\tilde{f}_{sh}$ and $\tilde{n}_{sh}$ are the RMS values of the shaking force and shaking moment. Considering the lower and upper limits on the link masses and their mass center locations, the problem of mechanism balancing is finally stated as

$$\text{Minimize } z(\mathbf{x}) = w_1 \tilde{f}_{sh} + w_2 \tilde{n}_{sh} \qquad (6.15a)$$

$$\text{Subject to } m_{i,\min} \leq \sum_{j=1}^{7} m_{ij} \leq m_{i,\max} \qquad (6.15b)$$

$$r_{i,\min} \leq r_i \leq r_{i,\max} \qquad (6.15c)$$

for $i=1$, ..., n, whereas $m_{i,\min}$ and $m_{i,\max}$, and $r_{i,\min}$ and $r_{i,\max}$ are the minimum and maximum limits on the mass and its mass center location of the ith link. Note here that the minimum moments of inertia of the ith link depends only on $m_{i,\min}$. For example, $I_{ixx,\min}$ with respect to O_i in the link-fixed frame is given by

$$I_{ixx,min} = \sum_{j=1}^{7} m_{ij}(d_{ijy}^2 + d_{ijz}^2) = m_{i,min}[(a_{i,i+1y} + r_{ijy})^2 + (a_{i,i+1z} + r_{ijz})^2] \quad (6.16)$$

where $r_{ijx}, r_{ijy}, r_{ijz}$ are the components of the 3-vector $\mathbf{r}_{ij}$ given in Table 6.1. Moreover, $a_{i,i+1x}, a_{i,i+1y}, a_{i,i+1z}$ are the components of the link length vector, $\mathbf{a}_{i,i+1}$. Therefore, the moments of inertia of the links are governed by the bound chosen on the link masses, as $\mathbf{r}_{ij}$ and $\mathbf{a}_{i,i+1}$ are constants in the local coordinate frame. Hence, the optimization problem finds a value of each point mass of each link while the total mass and its mass center location of each link are subjected to lower and upper limits. From the optimized values of point-masses m_{ij}^*, optimized total mass m_i^*, location of the mass center ($\bar{x}_i^*, \bar{y}_i^*, \bar{z}_i^*$), and inertia $I_{i,xx}^*, I_{i,yy}^*, I_{i,zz}^*, I_{i,xy}^*, I_{i,yz}^*, I_{i,zx}^*$ of each link are determined using the equimomental conditions, Eqs. (4.29-38).

Counterweight method

When the given unbalanced mechanism has been kinematically synthesized, and the mass distribution of the links has been determined according to load bearing capacity, etc. the mechanism can be balanced by attaching counterweights to the moving links. Assume that the counterweight mass, m_i^b, is attached to the ith link at ($\bar{x}_i^b, \bar{y}_i^b, \bar{z}_i^b$), as shown in Fig. 6.3. The equimomental system of the resulting link is shown in Fig. 6.4, where it is assumed that point-masses, m_{ij}^b, are placed at the same locations where the point-masses of the original link, m_{ij}^o were located. Then the mass of the counterweight, m_i^b, its mass center location, $\mathbf{r}_i^b$, and its inertias, $I_{i,xx}^b, I_{i,yy}^b, I_{i,zz}^b, I_{i,xy}^b, I_{i,yz}^b$, and $I_{i,zx}^b$, can be obtained using the equimomental conditions, Eq. (4.29-38). Now, for a mechanism having n moving links, the $7n$-vector of the design variables, $\mathbf{x}^b$, is

$$\mathbf{x}^b \equiv [\mathbf{m}_1^{bT}, ..., \mathbf{m}_n^{bT}]^T \quad (6.17)$$

where the 7-vectors, $\mathbf{m}_i^b$, is as follows:

$$\mathbf{m}_i^b \equiv \begin{bmatrix} m_{i1}^b & m_{i2}^b & m_{i3}^b & m_{i4}^b & m_{i5}^b & m_{i6}^b & m_{i7}^b \end{bmatrix}^T$$

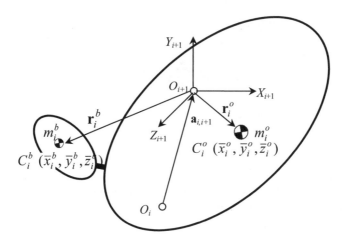

Fig. 6.3. The ith link with counterweight

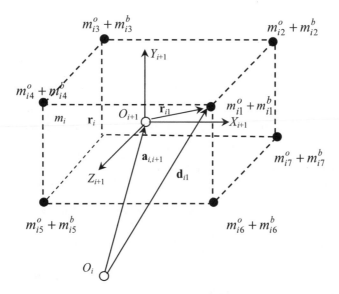

Fig. 6.4. Equimomental system of point-masses for the counterweighted ith link

Considering the lower and upper limits on the counterweight masses and their mass center locations the balancing problem is stated similar to Eq. (6.15) as

$$\text{Minimize } z(\mathbf{x}^b) = w_1 \tilde{f}_{sh} + w_2 \tilde{n}_{sh} \tag{6.18a}$$

$$\text{Subject to } m_{i,\min}^b \leq \sum_{j=1}^{7} m_{ij}^b \leq m_{i,\max}^b \tag{6.18b}$$

$$r_{i,\min}^b \leq r_i^b \leq r_{i,\max}^b \tag{6.18c}$$

for $i=1, \ldots, n$ and $j=1, \ldots, 7$.

The methodology for the optimization using either mass redistribution or counterweight is summarized below:

1. Normalize the given unbalanced mechanism as per Table 6.2.

Table 6.2. Normalization of link parameters

$a_{ij} =: \|\mathbf{a}_{ij}\| / a_m$	: Normalized distance between joints i and j
$d_i =: \|\mathbf{d}_i\| / a_m \; ; r_i =: \|\mathbf{r}_i\| / a_m$	: Normalized distance of the mass center
$m_i =: m_i / m_m^o$	: Normalized mass of the ith link
$\mathbf{I}_i =: \mathbf{I}_i / (m_m^o a_m^2)$	: Normalized inertia tensor of the ith link

a_m and m_m^o are the parameters of the reference link about which the mechanism is normalized

2. Given mass, its mass center location, and the inertia components of each link: m_i, $\bar{x}_i$, $\bar{y}_i$, $\bar{z}_i$, $I_{i,xx}$, $I_{i,yy}$, $I_{i,zz}$, $I_{i,xy}$, $I_{i,yz}$, $I_{i,zx}$ of the normalized unbalanced mechanism, find the set of point-mass parameters, i.e., m_{ij}, h_{ix}, h_{ix}, and h_{iz}, for the seven equimomental point-masses, $j=1, \ldots, 7$. These are obtained using Eqs. (4.42-45) and (4.55), where subscript i is to be used to denote the ith link.

3. Define the design variable vector, Eq. (6.13) for the mass redistribution and Eq. (6.17) for counterweight method.

4. Define the objective function and the constraints on link masses, as in Eq. (6.15) or Eq. (6.18), where the normalized shaking force and shaking moment are defined as $\bar{f} = \|\mathbf{f}_{sh}\| / (m_m^o a_m \omega_{in}^2)$ and $\bar{n} = \|\mathbf{n}_{sh}\| / (m_m^o a_m^2 \omega_{in}^2)$ with $\omega_{in} = 1$ rad/sec.

5. Solve the optimization problem posed in the above step, (4), using any standard software, say, optimization toolbox of MATLAB [126].

The optimization process is started with the parameters of the unbalanced mechanism as initial design vector.

6. In the case of mass redistribution method, the optimized total mass, m_i^*, its location, $(\bar{x}_i^*, \bar{y}_i^*, \bar{z}_i^*)$, and the inertias, $I_{i,xx}^*, I_{i,yy}^*, I_{i,zz}^*, I_{i,xy}^*, I_{i,yz}^*, I_{i,zx}^*$, of each link are obtained from the optimized values of the point-masses, m_{ij}^*, using the equimomental conditions, Eq. (4.29-30). Similarly, the optimized parameters for counterweights are determined in counterweight method.

7. Actual values of the link masses or counterweights mass, mass center location, and the components of the inertia tensor, are then obtained by multiplying with the normalizing factors, m_m^o, a_m, and $m_m^o a_m^2$, respectively.

6.2 Spatial RSSR Mechanism

The effectiveness of the proposed methodology to optimize the combined shaking force and shaking moment in a spatial mechanism is demonstrated here using the RSSR mechanism shown in Fig. 3.16 and its kinematically equivalent 7R mechanism given in Fig. 3.17. The shaking force and shaking moment with respect to the mid-point of O_1O_7 transmitted to the mechanism are obtained using Eqs. (6.11-12) as

$$\mathbf{f}_{sh} = -(\mathbf{f}_{01} + \mathbf{f}_{06}) \tag{6.19}$$

$$\mathbf{n}_{sh} = -(\mathbf{n}_{01} + \mathbf{n}_{06} + 0.5\tilde{\mathbf{a}}_{17}\mathbf{f}_{06} - 0.5\tilde{\mathbf{a}}_{17}\mathbf{f}_{01}) \tag{6.20}$$

Since only the links, 1, 3, and 6, are having some dimensions with some masses, their equimomental systems of point-masses are considered in the optimization process. Hence, the 21-vector, $\mathbf{x}$, of the design variables is as follows:

$$\mathbf{x} \equiv [\mathbf{m}_1^T, \mathbf{m}_3^T, \mathbf{m}_6^T]^T \qquad \text{for mass redistribution method}$$

$$\mathbf{x}^b \equiv [\mathbf{m}_1^{b^T}, \mathbf{m}_3^{b^T}, \mathbf{m}_6^{b^T}]^T \qquad \text{for counterweight method}$$

Moreover, it is essential to position the mass center of link 3 along O_3O_4 or X_4 so as to control the gyroscopic action on the mechanism [104,120]. Furthermore, it is considered that the minimum mass of each link is equal to its original mass, m_i^o, and the maximum can rise up to five times the original one. Optimization problem for the balancing of the 7R mechanism is then posed as

$$\text{Minimize } z = w_1 \tilde{f}_{sh} + w_2 \tilde{n}_{sh} \tag{6.21a}$$

For Mass redistribution

$$\text{Subject to } m_i^o \leq \sum_{j=1}^{7} m_{ij} \leq 5m_i^o \, ; \; r_i \leq a_i \tag{6.21b}$$

$$\sum_{j=1}^{7} m_{3j} [r_{3jy}]_4 = 0 \, ; \text{ and } \sum_{j=1}^{7} m_{3j} [r_{3jz}]_4 = 0 \tag{6.21c}$$

For Counterweight

$$\text{Subject to } 0 \leq \sum_{j=1}^{7} m_{ij}^b \leq 3m_i^o \, ; \; 0 \leq r_i^b \leq a_i \tag{6.21d}$$

$$\sum_{j=1}^{7} (m_{3j}^o + m_{3j}^b)[r_{3jx}]_4 = 0 \, ; \; \sum_{j=1}^{7} (m_{3j}^o + m_{3j}^b)[r_{3jy}]_4 = 0 \tag{6.21e}$$

for $i=1$, 3, 6, where a_i is length of the ith link. Moreover $[\mathbf{r}_{3j}]_4 \equiv [r_{3jx} \quad r_{3jy} \quad r_{3jz}]^T$ is the vector from O_4 to the point-mass, m_{3j}, in the frame $O_4 X_4 Y_4 Z_4$, which is fixed to link 3. The equalities, Eqs. (6.21c) and (6.21e), guarantee the location of the mass center of link 3 along the axis, X_4, of the local frame, i.e., $\mathbf{F}_4$.

The optimization toolbox of MATLAB [126] is used to solve the optimization problem of Eq. (6.21). Using "fmincon" function that is based on the Sequential Quadratic Programming (SQP) method [127] finds a minimum of the function, z of Eq. (6.21a). The whole algorithm is coded in the MATLAB to determine the time dependent behavior of the various relevant quantities, including the bearing forces and the driving torque.

Table 6.3 shows the normalized DH parameters, mass and inertia of the 7R mechanism. The mass center location and the elements of the inertia tensor of each link are given in their local coordinate frames. The equimomental point masses and their locations are obtained using Eqs. (4.42-45) and (4.55), which are given in Table 6.4. The results of the optimization problem, Eq. (6.21), are obtained for three sets of weighting factors, namely, (w_1, w_2)=(1.0, 0.0); (0.5, 0.5); (0.0, 1.0). Cases (1)-(3) corresponding to mass redistribution, whereas cases (4)-(6) represent counterweight balancing. Table 6.4 shows the optimum design vectors for all the cases. The geometry, mass and inertias for the normalized balanced mechanism corresponding to the cases are obtained using equimomental conditions and given in Table 6.5. Note that the conditions, Eqs. (6.21c) and (6.21e), is achieved and shown by bold-faced number in Table 6.6. Table 6.7 shows the comparison between the RMS values of the normalized dynamic quantities occurred during the motion cycle and those for the original mechanism, whereas Fig. 6.6 shows the comparison of the dynamic performances of the mechanism. Locations of link mass centers of the balanced normalized mechanism for case (2) are shown in Fig. 6.7. It is evident from Table 6.7 and Fig. 6.6 that a significant improvement in performances is achieved.

Table 6.3. DH parameters, and mass and inertia properties for the normalized 7R mechanism

Link i	a_i	b_i	α_i	θ_i	m_i^o	r_{ix}^o	r_{iy}^o	r_{iz}^o	I_{ixx}^o	I_{iyy}^o	I_{izz}^o
1	1.00	0.55	90	θ_1	1.000	-0.50	0	0	0.0017	0.3477	0.3477
2	0	0	90	θ_2	0	0	0	0	0	0	0
3	1.10	0	90	θ_3	1.093	-0.55	0	0	0.0018	0.4579	0.4579
4	0	0	90	θ_4	0	0	0	0	0	0	0
5	0	0	90	θ_5	0	0	0	0	0	0	0
6	0.50	0	90	θ_6	0.536	-0.25	0	0	0.0009	0.0489	0.0489
7	1.30	0.40	150	θ_7	-	-	-	-	-	-	-

Total normalized mass of the mechanism, $\sum m_i^o$ =2.629

Normalized with respect to a_1=0.1 m; m_1=0.084 kg; ω_{in}=1 rad/sec.

Table 6.4. Equimomental point masses of the normalized 7R mechanism

Link i	Point-masses							Distances		
	m_{i1}^o	m_{i2}^o	m_{i3}^o	m_{i4}^o	m_{i5}^o	m_{i6}^o	m_{i7}^o	h_{ix}	h_{iy}	h_{iz}
1	0.2500	-0.2122	0.4622	0	0.4622	-0.2122	0.2500	0.5889	0.0292	0.0292
3	0.2732	-0.2324	0.5057	0	0.5057	-0.2324	0.2732	0.6466	0.0287	0.0287
6	0.1340	-0.1114	0.2454	0	0.2454	-0.1114	0.1340	0.3007	0.0290	0.0290

Results of case (a), which is the shaking force balancing, are compared with the analytical conditions of Bagci [104]. Bagci obtained the analytical solution for the total force balance of the general RSSR linkage by redistributing the link masses. The method is based on making the mass center of the mechanism stationary. The conditions for the total force balance for the RSSR, as shown in Fig. 6.5, are as follows:

$$[r_{1z}]_2 = 0 \qquad\qquad (6.22a)$$

$$[r_{6y}]_7 = 0 \qquad\qquad (6.22b)$$

$$m_6[r_{6x}]_7 = m_3 a_6 (1 - r_3/a_3) \qquad\qquad (6.22c)$$

$$m_1([r_{1x}]_2 + a_1) = -m_3 a_1 r_3/a_3 \qquad\qquad (6.22d)$$

where $r_3 \equiv [r_{3x}]_4$, and $[r_{1y}]_2$ and $[r_{6y}]_7$ can be of any value. By specifying m_3 and r_3, one can determine the masses and the geometries of links 1 and 6 in the form of mass-distance product. This is the limitation of the analytical method.

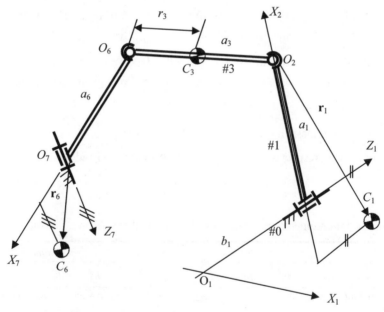

Fig. 6.5. Geometry of the full force balanced RSSR mechanism

However, for case (1), where complete shaking force balancing is achieved, the conditions, Eqs. (6.2), are calculated as follows:

$$[r_{1z}]_2 = 0.001$$

$$[r_{6y}]_7 = 0.000$$

$$m_6[r_{6x}]_7 = 0.5463; \qquad m_3 a_6 (1 - r_3 / a_3) = 0.5463$$

$$m_1([r_{1x}]_2 + a_1) = 0; \qquad -m_3 a_1 r_3 / a_3 = 0$$

where the data for the geometry and the masses of the links are taken from Table 6.6. Clearly the results satisfy the shaking force balancing conditions. Note that the analytical conditions are only four in twelve parameters, namely, m_i and the three components of vector, $\mathbf{r}_i$, for each link. Also, in the analytical conditions value of r_3 is undecided. No such restriction present in the proposed optimization methodology.

The following conclusions are accrued from the comparison of the results given in Tables 6.6 and 6.7.

- Total mass of the balanced mechanism is minimum when equal weights are applied to the shaking force and shaking moment, i.e., 3.199 for case (2) and 2.991 for case (5), for the mass redistribution and counterweight balancing, respectively.

- Total mass of the balanced mechanism in both the balancing methods are approximately same in the corresponding cases, e.g., 3.220 and 3.198 for cases (1) and (4), respectively.

- Both the shaking force and shaking moment are reduced more in the mass redistribution balancing than counterweight balancing. With the mass redistribution, reduction in the RMS values of 89% and 92% is achieved for the shaking force and shaking moment, respectively, over that of the original mechanism whereas with counterweight balancing 41% and 76% reductions are seen in the RMS values of the shaking force and shaking moment, respectively.

- It is observed that when shaking moment with respect to O_1 is taken, the reactions at O_7 are minimum and vise-versa. Hence, to incorporate effect of both f_{01} and f_{06}, the mid-point of O_1O_7 is chosen to determine the shaking moment.

Based on the above observations, the mass redistribution method is effective. However, when this is not possible, counterweight balancing needs to can be adopted.

Table 6.5. Design vectors for the normalized balanced mechanism

Cases (w_1, w_2)	Design vector

(a) Mass redistribution method

| (1) (1.0,0.0) | $[\;0.1212 \quad -0.3017 \quad 0.6463 \quad 0.1430 \quad 0.6053 \quad -0.3409 \quad 0.1606$
 $0.4466 \quad -0.1157 \quad 0.3313 \quad -0.1157 \quad 0.3306 \quad -0.1150 \quad 0.3309$
 $0.5285 \quad 0.2829 \quad -0.0096 \quad -0.2548 \quad -0.0976 \quad 0.1990 \quad 0.4442]^{\mathrm{T}}$ |

| (2) (0.5,0.5) | $[\;0.6521 \quad -0.7163 \quad -0.0964 \quad 0.4256 \quad 0.9352 \quad 0.2290 \quad -0.4292$
 $0.8384 \quad 1.1460 \quad -0.7195 \quad -0.7185 \quad 2.0459 \quad -1.6194 \quad 0.1199$
 $-0.6279 \quad 2.9295 \quad -1.7194 \quad 1.1138 \quad 0.2628 \quad -2.0301$
 $1.1777]^{\mathrm{T}}$ |

| (3) (0.0,1.0) | $[-1.5240 \quad -0.9882 \quad 0.9936 \quad -0.4674 \quad 0.6876 \quad 0.4905$
 $1.8079 \quad 0.3356 \quad 0.1637 \quad 0.1379 \quad -0.0907 \quad 0.9190 \quad -0.6174$
 $0.2449 \quad -0.1593 \quad 0.8132 \quad -0.3126 \quad 0.2651 \quad 0.5540 \quad -0.3610$
 $0.6984]^{\mathrm{T}}$ |

(b) Counterweight method

| (4) (1.0,0.0) | $[\;0.1716 \quad -0.1736 \quad 0.1026 \quad -0.0156 \quad 0.1825 \quad -0.0923 \quad 0.0246$
 $0.3383 \quad 0.0052 \quad 0.0120 \quad -0.1606 \quad 0.0120 \quad 0.0052 \quad 0.1777$
 $0.8146 \quad 0.6413 \quad -0.0413 \quad -0.1138 \quad -0.3768 \quad 0.3060 \quad 0.3780]^{\mathrm{T}}$ |

| (5) (0.5,0.5) | $[-1.1917 \quad -1.5375 \quad 0.4706 \quad -0.0625 \quad 0.6477 \quad 0.9568$
 $1.5094 \quad 0.1899 \quad -0.1244 \quad 0.0609 \quad -0.1264 \quad 0.0655 \quad -0.1290$
 $0.0635 \quad 0.2003 \quad 0.2840 \quad -0.1789 \quad -0.0969 \quad -0.0285 \quad 0.2999$
 $0.6255]^{\mathrm{T}}$ |

| (6) (0.0,1.0) | $[-2.8496 \quad -3.6776 \quad 3.5322 \quad 3.8820 \quad -6.2060 \quad 2.3213$
 $3.9066 \quad -0.8870 \quad 0.3400 \quad 0.1908 \quad 0.3576 \quad -0.5485 \quad 1.0792$
 $-0.5294 \quad -1.9386 \quad 3.5731 \quad 0.4023 \quad -1.4008 \quad 1.4840 \quad 0.6523$
 $-1.1643]^{\mathrm{T}}$ |

Table 6.6. Geometry, mass and inertia of the normalized balanced mechanism

(a) Mass redistribution method											
Case	Link i	m_i^*	r_{ix}^*	r_{iy}^*	r_{iz}^*	I_{ixx}^*	I_{iyy}^*	I_{izz}^*	I_{ixy}^*	I_{iyz}^*	I_{izx}^*
(1)	1	1.034	-1.000	0.005	0.001	0.0018	0.3594	0.3594	0.0032	0.0002	0.0031
	3	1.093	0.000	**0.000**	**0.000**	0.0018	0.4579	0.4579	-0.0043	-0.0002	-0.0043
	6	1.093	0.500	0.000	-0.009	0.0018	0.0997	0.0997	-0.0029	-0.0003	-0.0030
		3.220⁺									
(2)	1	1.000	-0.900	-0.014	0.108	0.0017	0.3477	0.3477	-0.0128	-0.0003	-0.0098
	3	1.093	-0.073	**0.000**	**0.000**	0.0018	0.4578	0.4578	-0.1293	0.0005	0.0760
	6	1.106	0.487	0.060	-0.096	0.0019	0.1010	0.1010	-0.0351	-0.0019	0.0860
		3.199									
(3)	1	1.000	-0.841	-0.145	-0.077	0.0017	0.3477	0.3477	0.0799	0.0012	0.0186
	3	1.093	-0.497	**0.000**	**0.000**	0.0018	0.4579	0.4579	-0.0324	0.0001	0.0256
	6	1.498	0.097	-0.006	-0.017	0.0025	0.1367	0.1367	-0.0080	-0.0001	0.0276
		3.591									

(b) Counterweight method											
Case	Link i	m_i^{b*}	r_{ix}^{b*}	r_{iy}^{b*}	r_{iz}^{b*}	I_{ixx}^{b*}	I_{iyy}^{b*}	I_{izz}^{b*}	I_{ixy}^{b*}	I_{iyz}^{b*}	I_{izx}^{b*}
(4)	1	0.200	-0.100	-0.004	0.043	0.0003	0.0695	0.0695	-0.0028	-0.0001	-0.0028
	3	0.390	1.100	**0.000**	**0.000**	0.0006	0.1633	0.1633	-0.0060	-0.0003	-0.0060
	6	1.608	0.500	0.018	-0.006	0.0027	0.1468	0.1468	-0.0048	-0.0005	-0.0048
		2.198⁺⁺									
(5)	1	0.793	-0.980	-0.200	-0.003	0.0014	0.2756	0.2756	0.0852	0.0002	0.0055
	3	0.000	0.000	**0.000**	**0.000**	0.0000	0.0000	0.0000	0.0000	0.0000	0.0000
	6	1.105	0.466	-0.018	-0.0093	0.0019	0.1009	0.1009	0.0017	-0.0003	0.0040
		1.898									
(6)	1	0.909	-0.977	0.028	-0.213	0.0015	0.3160	0.3160	0.4536	-0.0076	-0.0877
	3	0.003	0.695	**0.001**	**-0.001**	0.0000	0.0011	0.0011	0.0407	0.0017	-0.0142
	6	1.608	0.119	-0.006	-0.072	0.0027	0.1468	0.1468	-0.0404	0.0089	0.0294
		2.520									

⁺Total mass of the normalized mechanism, Σm_i^* for mass redistribution;

⁺⁺For counterweight method $m_i^* = m_i^o + m_i^{b*}$, Σm_i^* =3.198 ; Σm_i^* =2.991;

Σm_i^* =3.056 for cases (4)-(6), respectively.

Table 6.7. RMS values for the dynamic quantities

Case (w_1, w_2)	Input torque, $\tilde{\tau}$	Shaking force, $\tilde{f}_{sh}$	Shaking moment, $\tilde{n}_{sh}$ [*]
Original	0.1279	0.7772	0.7458
(1) (1.0, 0.0)	0.1102 (-14)	0.0050 (-99)	0.2904 (-61)
(2) (0.5, 0.5)	0.0940 (-27)	0.0875 (-89)	0.1119 (-92)
(3) (0.0, 1.0)	0.0322 (-75)	0.4078 (-48)	0.1797 (-85)
(4) (1.0, 0.0)	0.0699 (-45)	0.3051 (-61)	0.3520 (--53)
(5) (0.5, 0.5)	0.0639 (-50)	0.4614 (-41)	0.1785 (-76)
(6) (0.0, 10)	0.0562 (-56)	0.6451 (-17)	0.2920 (-61)

The value in the parentheses denotes the round-off percentage increase or decrease over the corresponding values for the original mechanism.

[*]Shaking moment with respect to the mid-point of the fixed link 7

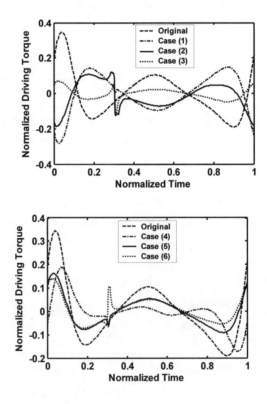

(a) Normalized driving torque

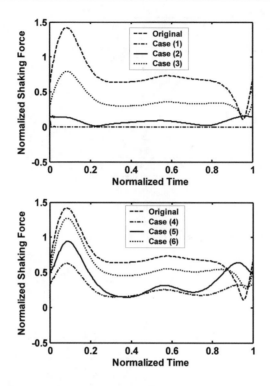

(b) Normalized shaking force

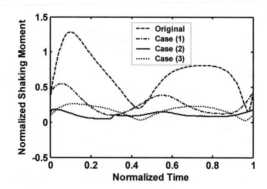

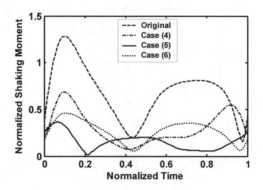

(c) Normalized shaking moment about the mid-point of the fixed link 7

Fig. 6.6. Dynamic performances of the 7R mechanism

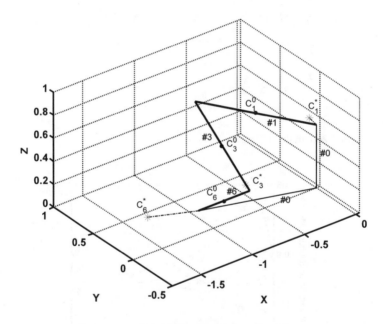

•: Original; *: Optimized location of link mass

Fig. 6.7. Locations of link mass centers of the balanced 7R mechanism, case (2)

6.3 Summary

In this chapter, balancing problem of spatial mechanisms is formulated as an optimization problem. The design variables and the constraints on them are identified by introducing the equimomental system of point-masses. Amongst several possibilities, the seven point-mass model is proposed, as explained in chapter 4. Using the equimomental point-masses, the equations of motion are reformulated to determine the shaking force, shaking moment, and other dynamic quantities. The spatial RSSR mechanism is balanced using the proposed methodology. The results depend upon the constraints like minimum and maximum masses and the objective function to be used to balance the mechanism. It is found that the objective function of combined shaking force and shaking moment with equal weights gives better results.

Appendix A: Coordinate Frames

In order to specify the configuration and dimension of a spatial mechanism, body-fixed coordinate frames are defined here. The notations used here are the modified version of those originally proposed by Denavit and Hartenberg [57] for the same purpose.

A.1 Denavit-Hartenberg Parameters

It is assumed that each link is coupled to its previous and the next one with one degree-of-freedom (DOF) joints, revolute or prismatic. For multi-DOF joints, they can be equivalently represented as a combination of the revolute and prismatic joints. Referring to Fig. A.1, the ith joint couples the $(i-1)$st link with the ith one.

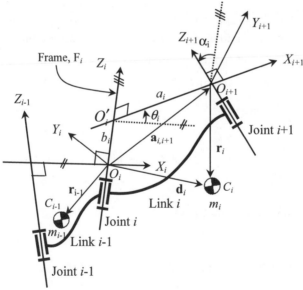

Fig. A.1. Coordinate frames and associated parameters

With each link, namely, the $(i\text{-}1)$st one, a Cartesian coordinate system, $O_iX_iY_iZ_i$, denoted by $\mathcal{F}_i$, is attached at O_i. Point O_i is the origin of the ith coordinate system and located at the intersection of X_i and Z_i. The axis, Z_i, is along the ith joint axis, whereas X_i is the common perpendicular between the two consecutive joint axes, Z_{i-1} and Z_i, directed from the former to the latter. The axis Y_i is then defined to complete the right-handed system, $O_iX_iY_iZ_i$. Based on the concept of the Devanit-Hartenberg (DH) conventions [57], the distance between Z_i and Z_{i+1} axes is defined as a_i, which is positive. The Z_i coordinate of the intersection of Z_i with X_{i+1} is denoted by b_i. Note that this quantity is a coordinate and, hence, can be either positive or negative. The angles between, Z_i and Z_{i+1}, and X_i and X_{i+1}, are α_i, and θ_i, and measured about the positive direction of X_{i+1} and Z_i, respectively. For a revolute joint, only θ_i varies and is called the joint variable, whereas for a prismatic joint, the joint variable is b_i. The dimensions and configuration of the linkage are then determined by a set of four parameters, a_i, b_i, α_i, and θ_i, which are referred as the DH parameters for each link that define its relative configuration within the system.

A.2 Transformations

The relative position and orientation between two bodies are fully specified by (i) a rotation matrix, $\mathbf{Q}_i$, carrying $\mathcal{F}_i$ into an orientation coincident with that of $\mathcal{F}_{i+1}$, and (ii) the position vector of the origin, O_{i+1}, relative to O_i. The orientation is obtained by two simple rotation matrices, θ_i about axis Z_i and α_i about X_{i+1}, i.e,

$$\mathbf{Q}_\theta \equiv \begin{bmatrix} C\theta_i & -S\theta_i & 0 \\ S\theta_i & C\theta_i & 0 \\ 0 & 0 & 1 \end{bmatrix} \text{ and } \mathbf{Q}_\alpha \equiv \begin{bmatrix} 1 & 0 & 0 \\ 0 & C\alpha_i & -S\alpha_i \\ 0 & S\alpha_i & C\alpha_i \end{bmatrix} \tag{A.1}$$

Hence, the 3×3 rotation matrix $\mathbf{Q}_i : \mathcal{F}_i \rightarrow \mathcal{F}_{i+1}$ is

$$\mathbf{Q}_i = \mathbf{Q}_\theta \mathbf{Q}_\alpha = \begin{bmatrix} C\theta_i & -S\theta_i C\alpha_i & S\theta_i S\alpha_i \\ S\theta_i & C\theta_i C\alpha_i & -C\theta_i S\alpha_i \\ 0 & S\alpha_i & C\alpha_i \end{bmatrix} \tag{A.2}$$

where $S\theta_i \equiv \sin\theta_i$, $C\theta_i \equiv \cos\theta_i$, $S\alpha_i \equiv \sin\alpha_i$, and $C\alpha_i \equiv \cos\alpha_i$. Knowing the orientation matrix, $\mathbf{Q}_i$, the representation of any three-dimensional Euclidean vector $\mathbf{r}$ in $\mathcal{F}_{i+1}$ can be transformed into that of $\mathcal{F}_i$ as:

$$[\mathbf{r}]_i = \mathbf{Q}_i[\mathbf{r}]_{i+1} \tag{A.3}$$

Moreover, the vector, $\mathbf{a}_{i,i+1}$, from O_i to O_{i+1}, as indicated in Fig. A.1, in terms of the DH parameters is as follows:

$$\mathbf{a}_{i,i+1} = b_i \mathbf{e}_i + a_i \mathbf{x}_{i+1} \tag{A.4}$$

where $\mathbf{e}_i$ and $\mathbf{x}_{i+1}$ are the unit vectors along Z_i and X_{i+1}, respectively, which have simple representation in $\mathcal{F}_i$ and $\mathcal{F}_{i+1}$ as:

$$[\mathbf{e}_i]_i = [0 \quad 0 \quad 1]^T \text{ and } [\mathbf{x}_{i+1}]_{i+1} = [1 \quad 0 \quad 0]^T \tag{A.5}$$

Similarly, any 3×3 matrix $\mathbf{M}$ in $\mathcal{F}_{i+1}$ is transformed into that of $\mathcal{F}_i$ [68] as:

$$[\mathbf{M}]_i = \mathbf{Q}_i[\mathbf{M}]_{i+1}\mathbf{Q}_i^T \tag{A.6}$$

The inverse relations for a vector and matrix follow immediately from Eqs. (A.3) and (A.6) as

$$[\mathbf{r}]_{i+1} = \mathbf{Q}_i^T[\mathbf{r}]_i \text{ and } [\mathbf{M}]_{i+1} = \mathbf{Q}_i^T[\mathbf{M}]_i\mathbf{Q}_i \tag{A.7}$$

A.3 Inertia Tensor

Inertia tensor of a rigid body having volume v with respect to a point O is defined [10] as

$$\mathbf{I}_o = -\int_v \tilde{\mathbf{r}}\tilde{\mathbf{r}}\,dm \tag{A.8}$$

where $\mathbf{r}$ is the vector from O to any point on the body, and $\tilde{\mathbf{r}}$ is the cross-product tensor associated with vector $\mathbf{r}$, which is defined similar to Eq. (2.5). Using Eq. (A.6) and the property, $\mathbf{Q}_i^T\mathbf{Q}_i = \mathbf{1}$, it can be shown

$$[\mathbf{I}_o]_i = \mathbf{Q}_i[\mathbf{I}_o]_{i+1}\mathbf{Q}_i^T \tag{A.9}$$

Appendix B: Topology Representation

A closed-loop system can be converted into an equivalent spanning-tree by cutting its appropriate joints. The appropriate joints to be cut can be identified using the concept of cumulative degree of freedom (CDOF) in graph theory [121]. The total degree of freedom of all the joints lie between the base link and a link of the system along any path is called the CDOF. For the four-bar example shown in Fig. B.1, the CDOF between the coupler, #2, and the fixed base, #0, is 2 along both the paths, i.e., 0-1-2 and 0-3-2, whereas those of between link, #1, and the base, #0, are 1 and 3 along the paths 0-1 and 0-3-2-1, respectively.

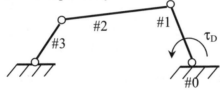

Fig. B.1. Four-bar mechanism

The joint to be cut in a closed-loop of a multiloop mechanism is decided based on the path whose CDOF is minimum. For the four-bar mechanism cutting of joint between links, #1 and #2 or #2 and #3, gives the minimum CDOF, i.e., 1 and 2.

In order to illustrate the concept for a multiloop mechanism, consider the scrapping mechanism of the carpet scrapping mechanism shown in Fig 4.5(b), which is reproduced here in Fig. B.2(a). Note that links, #1, #3, and #4, are connected to the base link, #0. Links, #5 and #7, are not directly connected to each other. They are both joined to #2. Moreover, all the joints between the links are revolute. The directed graph is now shown in Fig. B.2(b), where the node representing the base link, #0, is drawn first. In the second step, nodes of links, #1, #3, and #4, connecting link #0 are drawn, and connected with the edges, 1, 2, and 3, respectively. In the next step, the nodes representing the remaining links, #2, #5, and #6, are drawn, followed by the edges for the joints between them. As a result, edges 5, 6, and 7 close the loops, #0-#1-#2-#3-#0, #0-#1-#2-#5-#4-#0, and #0-#1-#2-

#7-#6-#4-#0, respectively. These edges are to be chosen for cutting the closed-loops. Cutting of other edges of the loops will lead to the same or higher CDOF. For example, the cutting of the edge, 5, in the loop, #0-#1-#2-#3-#0, gives CDOF equal to 2 and 1 in the paths, 0-1-2 and 0-3, respectively, whereas the cutting the edge, 2, gives the CDOF equal to 3 in the path 0-1-2-3-0. Hence, the edge, 2, is certainly not the candidate for the cut of the loop under consideration.

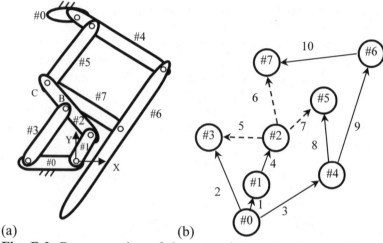

(a)   (b)

Fig. B.2. Representation of the scrapping mechanism (a) The scrapping mechanism (b) Its graphical representation

References

1. Eberhard P, Schiehlen W (2006) Computational dynamics of multibody systems: history, formalisms, and applications. ASME Journal of Computational and Nonlinear Dynamics, 1(1):3-12
2. Schiehlen W (1997) Multibody System Dynamics: Roots and Perspectives. Multibody Systems Dynamics 1:149-188
3. Goldstein, H., Poole, C., and Safko, J., 2002, *Classical Mechanics*, 3rd, Pearson Education, (Singapore) Pte. Ltd., Delhi.
4. Greenwood DT (1988) Principles of Dynamics. Prentice-Hall of India Pvt. Ltd., New Delhi
5. Schiehlen W (1990) Multibody Systems Handbook. Springer-Verlag, Berlin Heidelberg
6. Roberson RE, Schwertassek R (1988) Dynamics of multibody systems. Springer-Verlag, Berlin Heidelberg
7. Paul B (1975) Analytical dynamics of mechanisms-A computer oriented overview. Mechanism and Machine Theory 10(6):481-507
8. Fuhrer C, Schwertassek R (1990) Generation and solution of multibody system equations. Int. J. Non-Linear Mechanics 25(2/3):127-141
9. Gupta VK (1974) Dynamic analysis of multi-rigid-body systems. ASME Journal of Engineering for Industry 96(3):886-892
10. Nikravesh PE (1988) Computer-aided analysis of mechanical systems. Prentice-Hall, Englewood Cliffs, New Jersey
11. Shabana AA (2005) Dynamics of multibody systems, Cambridge University Press, New York
12. Seth PN, Uicker JAJ (1972) IMP (Integrated Mechanism Program): a computer-aided design analysis system for mechanisms and linkages. ASME Journal of Engineering for Industry 94(2):454-464
13. Orlando N, Chance MA, Calahan DA (1977) A sparsity oriented approach to the dynamic analysis and design of mechanical systems, Part I and II. Journal of Engineering for Industry 99(3):773-784
14. Uicker Jr. JJ (1967) Dynamic force analysis of spatial linkages. ASME Journal of Applied Mechanics 34:418-424
15. Smith DA, Chance MA, Rubines AC (1973) The automatic generation of a mechanical model for machinery. ASME Journal of Engineering for Industry 95(2):629-635
16. Chance MA, Bayazitoglu YO (1971) Development and application of a generalized d'alembert force for multifreedom mechanical systems. ASME Journal of Engineering for Industry 93(1):317-327

17. Chance MA (1967) Analysis of the time-dependent of multi-freedom mechanical systems in relative coordinates. ASME Journal of Engineering for Industry 89(1):119-125
18. Smith DA (1973) Reaction force analysis in generalized machine systems. ASME Journal of Engineering for Industry 95(2):617-623
19. Milner JR, Smith DA (1979) Topological reaction force analysis. ASME Journal of Mechanical Design 101(2):192-198
20. Kane TR, Levinson DA (1983) Multibody dynamics. ASME Journal of Applied Mechanics 50(4):1071-1078
21. Kane TR, Levinson DA (1985) Dynamics: Theory and Applications. McGraw-Hill Book Company, New York.
22. Kamman JW, Huston RL (1984) Dynamics of constrained multibody systems. ASME Journal of Applied Mechanics 51(4):899-903
23. Kamman J W, Huston RL (1984) Constrained multibody system dynamics: An automated approach. Computers & Structures 18(6):999-1003
24. Wang JT, Huston RL (1988) Computational methods in constrained multibody dynamics: Matrix formalisms. Computers & Structures 29(2):331-338
25. Huston RL, Liu CQ (2005) Advances in computational methods for multibody system dynamics. Computer Modeling in Engineering & Science 10(2):143-152.
26. Nikravesh PE (1984) Some methods for dynamic analysis of constrained mechanical systems: A survey. In: Haug E J (Ed.) Proc. of NATO ASI, 9, Computer Aided Analysis and Optimization of Mechanical System Dynamics, Springer-Verlag Berlin Heidelberg, pp. 351-368
27. Baumgarte J (1972) Stabilization of constraints and integrals of motion. Computer methods in Applied Mechanics and Engineering 1:1-16
28. Yu Q, Chen IM (2000) A direct violation correction method in numerical simulation of constrained multibody systems. Computational Mechanics 26(1):52-57
29. Lin ST, Haung J (2002) Numerical integration of multibody mechanical systems using baumgarte's constraint stabilization method. Journal of the Chinese Institute of Engineers 25(2):243-252
30. Tseng FC, Ma ZD, Hulbert GM (2003) Efficient numerical solution of constrained multibody dynamics systems. Computer Methods in Applied Mechanics and Engineering 192(3-4):439-472
31. Yen J, Petzold LR (1998) An efficient Newton-type iteration for the numerical solution of highly oscillatory constrained multibody dynamic systems. SIAM Journal on Scientific Computing 19(5):1513-1534
32. Wehage RA, Haug EJ (1982) Generalized coordinate partitioning for dimension reduction in analysis of constrained dynamic systems. ASME Journal of Mechanical Design 104:247-255
33. Huston RL, Passerello CE (1974) On constraint equations—A new approach. ASME Journal of Applied Mechanics 41:130-1131
34. Hemami H, Weimer FC (1981) Modeling of nonholonomic dynamic systems with applications. ASME Journal of Applied Mechanics 48(1):177-182

35. Singh RP, Likins PW (1985) Singular value decomposition for constrained dynamical systems. ASME Journal of Applied Mechanics 52(4):943-948
36. Mani NK, Haug EJ, Atkinson KE (1985) Application of singular value decomposition for analysis of mechanical system dynamics. ASME Journal of Mechanisms, Transmissions, and Automation in Design 107(1):82-87
37. Amirouche FML, Jia T, and Ider SK (1988) A recursive householder transformation for complex dynamical systems with constraints. ASME Journal of Applied Mechanics 55(3):729-734
38. Kim SS, Vanderploeg MJ (1986) QR decomposition for state space representation of constrained mechanical dynamic systems. ASME Journal of Mechanisms, Transmissions, and Automation in Design 108(2):183-188
39. Liang CG, Lance GM (1987) A differentiable null space method for constrained dynamic analysis. ASME Journal of Mechanisms, Transmissions, and Automation in Design 109(3):405-411
40. Wamper C, Buffinton K, Shu-hui J (1985) Formulation of equations of motion for systems subject to constraints. ASME Journal of Applied Mechanics 52(2):465-470
41. Blajer W, Bestle D, Schiehlen W (1994) An orthogonal complement matrix formulation for constrained multibody systems. ASME Journal of Mechanical Design 116(2):423-428
42. Blajer W (1992) A projection approach to constrained dynamic analysis. ASME Journal of Applied Mechanics 59(3):643-649.
43. Blajer W (2001) A geometrical interpretation and uniform matrix formulation of multibody system dynamics. Zeitschrift fur Angewandte Mathematik und Mechanik (ZAMM) 81(4):247-259
44. Blajer W (2004) On the determination of joint reactions in multibody mechanisms. ASME Journal of Mechanical Design 126(2):341-350
45. Hui QZ, Bin W, Min F. Guan (1994) A new approach to formulation of the governing equations of constrained multibody systems. Computers & Structures, 53(4):1001-1006
46. Stepanenko Y, Vukobratovic M (1976) Dynamics of articulated open-chain active mechanisms. Mathematical Biosciences 28(1-2):137-170
47. Luh JYS, Walker MW, Paul RPC (1980) On-Line computational scheme for mechanical manipulators. ASME Journal of Dynamic Systems, Measurement and Control 102(2):69-76
48. Hollerbach JM (1980) A recursive lagragian formulation of manipulator dynamics and a comparative study of dynamics formulation complexity. IEEE trans. Systematic man Cybernetics 10(11):730-736
49. Silver WM (1982) On the equivalence of lagrangian and newton-euler dynamics for manipulators. Int. journal of Robotics Research 1(2):60-70
50. Walker MW, Orin DE (1982) Efficient dynamic computer simulation of robotic mechanisms. ASME Journal of Dynamics Systems, Measurement and Control 104:205-211

51. Jain A (1991) Unified formulation of dynamics for serial rigid multibody systems. Journal of Guidance, Control and Dynamics 14((3):531-542
52. Featherstone R (1987) Robot dynamics algorithms, Kluwer Academic Publishers
53. Bae DS, Haug EJ (1987) A recursive formulation for constrained mechanical system dynamics: Part I. Open systems. Int. J. of Mechanics of Structures and Machines 15(3):359-382
54. Bae DS, Haug EJ (1987) A recursive formulation for constrained mechanical system dynamics: Part II. Closed loop systems. Int. J. of Mechanics of Structures and Machines 15(3):481-506
55. Rodriguez G, Jain A, Kreutz-Delgado K (1991) A spatial operator algebra for manipulator modeling and control. Int. Journal of Robotics Research 10(4):371-381
56. Stelzle W, Kecskemethy A, Hiller M (1995) A comparative study of recursive methods. Archive of Applied Mechanics 66(1-2):9-19
57. Denavit J, Hartenberg RS (1955) A kinematic notation for lower-pair mechanisms based on matrices. ASME Journal of Applied Mechanics 77:215-221
58. Angeles J, Ma O (1988) Dynamic simulation of n-axis serial robotic manipulators using a natural orthogonal complement. International Journal of Robotic Research 7(5):32-47
59. Angeles J, Ma O, Rojas A (1989) An algorithm for the inverse dynamics of n-axis general manipulators using Kane's equations. Computers and Mathematics with Applications 17(12):1545-1561
60. Saha SK (1999) Dynamics of serial multibody systems using the decoupled natural orthogonal complement matrices. ASME Journal of Applied Mechanics, 66(4):986-996
61. Saha SK, Schiehlen WO (2001) Recursive kinematics and dynamics for closed loop multibody systems. Int. J. of Mechanics of Structures and Machines 29(2):143-175
62. Sohl GA, Bobrow JE (2001) A recursive multibody dynamics and sensitivity algorithm for branched kinematic chains. ASME Journal of Dynamics, Measurement, and Control 123(3):391-399
63. Balafoutis CA (1994) A survey of efficient computational methods for manipulator inverse dynamics. Journal of Intelligent and Robotic Systems 9(1-2):45-71
64. Anderson KS, Critchley JH (2003) Improved order-n performance algorithm for the simulation of constrained multi-rigid-body dynamic systems. Multibody System Dynamics 9(2):185-212
65. Critchley JH, Anderson KS (2003) A generalized recursive coordinate reduction method for multibody system dynamics. Int. Journal for Multiscale Computational Engineering 1(2&3):181-199
66. Jerkovskey W (1978) The structure of multibody dynamics equations. Journal of Guidance and Control 1(3):173-182
67. Kim SS, Vanderploeg MJ (1986) A general and efficient method for dynamic analysis of mechanical systems using velocity transformation. ASME Journal of Mechanisms, Transmissions, and Automation in Design 108(2):176-182

68. Angeles J (1997) Fundamental of robotic mechanical systems: theory, methods, and algorithms. Spring-Verlag New York

69. Angeles J, Lee S (1988) The formulation of dynamical equations of holonomic mechanical systems using a natural orthogonal complement. ASME Journal of Applied Mechanics 55(1):243-244.

70. Keat JE (1990) Multibody system order n dynamics formulation based on velocity transform method. Journal of Guidance, Control & Dynamics 13(2):207-212

71. Nikravesh PE (1990) Systematic reduction of multibody equations of motion to a minimal set. Int. J. Non-Linear Mechanics 25(2/3):143-151

72. Nikravesh PE, Gim G (1993) Systematic construction of the equations of motion for multibody systems containing closed kinematic loops. ASME Journal of Mechanical Design 115(1):143-149

73. Gacia DE Jalon J, Bayo E (1994) Kinematic and dynamic simulation of multibody systems-The real-time challenge. Springer-Verlag, New York

74. Gracia De Jalon J., Unda J., and Avello, A., 1986, "Natural Coordinates for the Computer Analysis of Multibody Systems," Computer Methods in Applied Mechanics and Engineering, 56(3), pp. 309-327.

75. Jimenez JM, Avello AN, Gracia De Jalon J, Avello AL (1995) An efficient implementation of the velocity transformation method for real-time dynamics with illustrative examples. In: Pereira MFOS, Ambrosio JAC (eds.) Computational Dynamics in Multibody Systems, Kluwer Academic Publishers, London, pp. 15-35

76. Cuadrado J, Dopico D, Gonzalez, Naya M A (2004) A combined penalty and recursive real-time formulation for multibody dynamics. ASME Journal of Mechanical Design, ASME Journal of Mechanical Design 126(4):602-608

77. Attia H A (2003) A matrix formulation for the dynamic analysis of spatial mechanisms using point coordinates and velocity transformation. Acta Mechanica 165:207-222

78. Saha SK, Prasad R, Mandal AK (2003) Use of hoeken's and pantograph mechanisms for carpet scrapping operations. In: Saha SK (ed) Proc. of 11th Nat. Conf. On Machines and Mechanisms, Dec. 18-19, IIT Delhi, pp. 732-738

79. Saha SK (2003) Simulation of industrial manipulator based on the UDUT decomposition of inertia matrix. Multibody System Dynamics 9:63-85

80. Duffy J (1978) Displacement analysis of the generalized RSSR mechanism. Mechanism and Machine Theory 13:533-541

81. Eber-Uphoff I, Gosselin MC, Laliberte T (2000) Static balancing of spatial parallel platform mechanisms-Revisited. ASME Journal of Mechanical Design 122(1):43-51

82. Quang PR, Zhang WJ (2005) Force balancing of robotic mechanisms based on adusmenet of kinematic parameters. ASME Journal of Mechanical Design, 127(3):433-440

83. Lowen GG, Tepper FR, Berkof RS (1983) Balancing of linkages-an updates. Mechanisms and Machine Theory 18(3):213-220

84. Shchepetil'nikov VA (1968) The determination of the mass centers of mechanisms in connection with the problem of mechanism balancing. Journal of Mechanism 3:367-389

85. Berkof RS, Lowen GG (1969) A new method for completely force balancing simple linkages. ASME Journal of Engineering for Industry 91(1):21-26

86. Kosav IS (1988) A new general method for full force balancing of planar linkages. Mechanism and Machine Theory 23(6):475-480

87. Lowen GG, Tepper FR, Berkof RS (1974) The quantitative influence of complete force balancing on the forces and moments of certain families of four-bar linkages, Mechanism and Machine Theory 9:299-323

88. Elliott JL, Tesar D (1977) The theory of torque, shaking force, and shaking moment balancing of four link mechanisms. ASME Journal of Engineering for Industry 99(3):715-722

89. Kamenskii VA (1968) On the questions of the balancing of plane linkages. Journal of Mechanism 3:303-322

90. Tricamo SJ, Lowen GG (1983) A novel method for prescribing the maximum shaking force of a four-bar linkage with flexibility in counterweight design. ASME Journal of Mechanisms, Transmission, and Automation in Design 105:511-519

91. Tricamo SJ, Lowen GG (1983) Simultaneous optimization of dynamic reactions of a four-bar linkage with prescribed maximum shaking force. ASME Journal of Mechanisms, Transmission, and Automation in Design 105:520-525

92. Berkof RS (1973) Complete force and moment balancing of inline four-bar linkage. Mechanism and Machine Theory 8:397-410

93. Esat I, Bahai H (1999) A theory of complete force and moment balancing of planar linkage mechanisms. Mechanism and Machine Theory 34:903-922

94. Ye Z, Smith MR (1994) Complete balancing of planar linkages by an equivalent method. Mechanism and Machine Theory 29(5):701-712

95. Arakelian VH, Smith MR (1999) Complete shaking force and shaking moment balancing of linkages. Mechanism and Machine Theory 34:1141-1153

96. Koshav IS (2000) General theory of complete shaking moment balancing of planar linkages: A critical review. Mechanism and Machine Theory 35:1501-1514

97. Arakelian VH, Smith MR (2005) Shaking force and shaking moment balancing of mechanisms: A historical review with new examples. ASME Journal of Mechanical Design 127:334-339

98. Berkof RS, Lowen GG (1971) Theory of shaking moment optimization of forced-balanced four-bar linkages. ASME Journal of Engineering for Industry 93B(1):53-60.

99. Carson WL, Stephenes JM (1978) Feasible parameter design spaces for force and root-mean-square moment balancing an in-line 4r 4-bar synthesized for kinematic criteria. Mechanism and Machine Theory 13:649-658.

100.Hains RS (1981) Minimum RMS shaking moment or driving torque of a force-balanced linkage using feasible counterweights. Mechanism and Machine Theory 16:185-190

101. Arakelian V, Dahan M (2001) Partial shaking moment balancing of fully force balanced linkages. Mechanism and Machine Theory 36:1241-1252.

102. Wiederrich JL, Roth B (1976) Momentum balancing of four-bar linkages. ASME Journal of Engineering for Industry 98(4):1289-1285

103. Kuafman RE, Sandor GN (1971) Complete force balancing of spatial linkages. ASME Journal of Engineering for Industry 93:620-626

104. Bagci C (1983) Complete balancing of space mechanisms-shaking force balancing. ASME Journal of Mechanisms, Transmissions, and Automation in Design 105:609-616.

105. Ning-Xin Chen (1984a) The complete shaking force balancing of a spatial linkage. Mechanism and Machine Theory 19(2):243-255

106. Ning-Xin Chen (1984b) Partial balancing of the shaking force of a spatial 4-bar RCCC linkage by the optimization method. Mechanism and Machine Theory 19(2):257-265

107. Yue-Qing Y (1987) Optimum shaking force and shaking moment balancing of the RSSR spatial linkage. Mechanism and Machine Theory 22(1):39-45

108. Yue-Qing Y (1987) Research on complete shaking force and shaking moment balancing of spatial linkages. Mechanism and Machine Theory 22(1):27-37

109. Chiou ST, Shieh MG, Tsai RJ (1992) The two rotating-mass balancers for partial balancing of spatial mechanisms. Mechanism and Machine Theory, 32(5):617-628

110. Routh EJ (1905) Treatise on the dynamics of a system of rigid bodies. Elementary Part I, Dover Publication Inc., New York

111. Wenglarz RA, Forarasy AA, Maunder L (1969) Simplified dynamic models. Engineering 208:194-195

112. Huang NC (1983) Equimomental system of rigidly connected equal particles. Journal of Guidance, Control and Dynamics 16(6):1194-1196

113. Sherwood AA, Hockey BA (1969) The optimization of mass distribution in mechanisms using dynamically similar systems. Journal of Mechanism 4: 243-260

114. Hockey BA (1972) The minimization of the fluctuation of input-shaft torque in plane mechanisms. Mechanism and Machine Theory 7:335-346

115. Lee TW, Cheng C (1984) Optimum balancing of combined shaking force, shaking moment, and torque fluctuations in high speed linkages. ASME Journal of Mechanisms, Transmissions, and Automation in Design 106(2):242-251

116. Qi NM, Pennestri (1991) Optimum balancing of four-bar linkages a refined algorithm. Mechanism and Machine Theory 26(3):337-348

117. Molian S (1973) Kinematics and dynamics of the RSSR mechanism. Mechanism and Machine Theory 8:271-282

118. Gill GS, Freudenstein F (1983) Minimization of inertia-induced forces in spherical four-bar mechanisms. Part 1: The general spherical four-bar linkage. ASME Journal of Mechanisms, Transmissions, and Automation in Design 105:471-477

119. Rahman S (1996) Reduction of inertia-induced forces in a generalized spatial mechanism. Ph.D. Thesis, Dept. of Mech. Eng., The New Jersey Institute of

Technology http://www.library.njit.edu/etd/1990s/1996/njit-etd1996-017/njit-etd1996-017.html

120. Feng B, Morita N, Torii T, Yoshida S (2000) Optimum balancing of shaking force and shaking moment for spatial rssr mechanism using genetic algorithm. JSME International Journal 43(3):691-696

121. Deo N (1974) Graph theory with application in engineering and computer science, Prentice-Hall, Englewood Cliffs NJ

122. Strang G (1998) Linear algebra and its applications. Harcourt Brace Jovanovich Publisher, Florida

123. MSC.ADAMS (Automatic Dynamic Analysis of Mechanical Systems) Version 2005.0.0 Jul 22, 2004

124. McPhee JJ (1996) On the use of linear graph theory in multibody system dynamics. Nonlinear Dynamics 9:73-90

125. Conte FL, George GR, Mayne RW, Sadler JP (1975) Optimum mechanism design combining kinematic and dynamic-force considerations. ASME Journal of Engineering for Industry 95(2):662-670

126. MATLAB (2004) Optimization Toolbox, Version 7.0.0.19920 (R14)

127. Arora JS (1989) Introduction to optimum design. McGraw-Hill Book Company, Singapore.

128. Yan H, Soong R (2001) Kinematic and dynamic design of four-bar linkages by links counterweighing with variable input speed. Mechanism and Machine Theory 36:1051-1071

129. Lowen GG, Berkof RS (1968) Survey of investigations into the balancing of linkages. Mechanism 3:221-231

130. Khan WA, Krovi VN, Saha SK, Angeles J (2005) Modular and recursive kinematics and dynamics for parallel manipulators. Multibody System Dynamics 14(3-4):19-55

131. Khan WA, Krovi VN, Saha SK, Angeles J (2005) Recursive kinematics and inverse dynamics for a planar 3R parallel manipulator. ASME Journal of Dynamic Systems, Measurement and Control 27(4):529-536

132. Rodriguez G, Jain A, Kreutz-Delgado K (1992) Spatial operator algebra for multibody system dynamics. The Journal of Astronautical Sciences 40(1):27-50

133. Norton R (1992) Design of machinery: an introduction to synthesis and analysis of mechanisms and machines, McGraw-Hill, New York

134. Chaudhary H, (2006) Dynamic analysis and minimization of inertia forces in mechanisms, Ph.D. thesis, Department of Mechanical Engineering, IIT Delhi.

Index

Printing: Krips bv, Meppel, The Netherlands
Binding: Stürtz, Würzburg, Germany